AF896783

Green Energy and Technology

Climate change, environmental impact and the limited natural resources urge scientific research and novel technical solutions. The monograph series Green Energy and Technology serves as a publishing platform for scientific and technological approaches to "green"—i.e. environmentally friendly and sustainable—technologies. While a focus lies on energy and power supply, it also covers "green" solutions in industrial engineering and engineering design. Green Energy and Technology addresses researchers, advanced students, technical consultants as well as decision makers in industries and politics. Hence, the level of presentation spans from instructional to highly technical.

Indexed in Scopus.

More information about this series at http://www.springer.com/series/8059

Pratap Bhattacharyya · Himanshu Pathak ·
Sharmistha Pal

Climate Smart Agriculture

Concepts, Challenges, and Opportunities

Springer

Pratap Bhattacharyya
ICAR-National Rice
Research Institute (NRRI)
Cuttack, India

Himanshu Pathak
ICAR-National Institute
of Abiotic Stress Management
Baramati, Maharashtra, India

Sharmistha Pal
ICAR-Indian Institute
of Soil and Water Conservation
Research Centre
Chandigarh, India

ISSN 1865-3529 ISSN 1865-3537 (electronic)
Green Energy and Technology
ISBN 978-981-15-9131-0 ISBN 978-981-15-9132-7 (eBook)
https://doi.org/10.1007/978-981-15-9132-7

© The Editor(s) (if applicable) and The Author(s), under exclusive license to Springer Nature Singapore Pte Ltd. 2020
This work is subject to copyright. All rights are solely and exclusively licensed by the Publisher, whether the whole or part of the material is concerned, specifically the rights of translation, reprinting, reuse of illustrations, recitation, broadcasting, reproduction on microfilms or in any other physical way, and transmission or information storage and retrieval, electronic adaptation, computer software, or by similar or dissimilar methodology now known or hereafter developed.
The use of general descriptive names, registered names, trademarks, service marks, etc. in this publication does not imply, even in the absence of a specific statement, that such names are exempt from the relevant protective laws and regulations and therefore free for general use.
The publisher, the authors and the editors are safe to assume that the advice and information in this book are believed to be true and accurate at the date of publication. Neither the publisher nor the authors or the editors give a warranty, expressed or implied, with respect to the material contained herein or for any errors or omissions that may have been made. The publisher remains neutral with regard to jurisdictional claims in published maps and institutional affiliations.

This Springer imprint is published by the registered company Springer Nature Singapore Pte Ltd.
The registered company address is: 152 Beach Road, #21-01/04 Gateway East, Singapore 189721, Singapore

Foreword

The World population would be touching around 10 Billion mark by 2050 according to the United Nations and to feed such a vast population there is a need to produce 60% more food. This is to be achieved in a situation when natural resources are depleting, and climate change is posing serious threat on agricultural systems. The increasing constraints of freshwater availability for crop and livestock are also a matter of concern. Therefore, the world scientific community is seriously concerned about the current situation which is regularly discussed in various climate summits. There is a need to address both opportunity and constraints on food security, system resilience, adaptation, and mitigation with synergy. The "climate-smart agriculture" (CSA) is a holistic solution of the above-mentioned complex problem. The encouragement must be given on "technological innovation," "investment," and "site-specific action" to "climate-smart agriculture" to address the issues of climate change and food security jointly.

The CSA has three pillars, viz., "sustainable food security," "system resilience and adaptation," and "mitigation to climate change." It is a powerful concept that directly focuses on agriculture and climate change nexus with location-specific flexibilities. Therefore, the CSA is a triple win for farmers/producers, investors, and consumers without regrets, trade-offs, and losers. However, the broad political will, efficient implementation capacity and enough financial commitment is necessary for promoting CSA. The present government and ICAR are providing considerable efforts to promote CSA through farmers' participation. This book seeks to provide a broad coverage of the topic with in-depth analysis that this complex issue demands and deserves. The chapters cover most of the aspects of CSA starting from its importance, dimensions, management practices, barriers to adaptation, financing as well as policy issues for mainstreaming it to national core policy. These important issues would generate comprehensive response from researchers, students, policy-makers, and public at a large about the climate change and its mitigation to benefit the Indian agriculture.

I appreciate the authors for presenting the different issues, challenges, and opportunities of "climate-smart agriculture" with a lucid and comprehensive way providing small flowcharts and easy explanations. I compliment the authors for their painstaking efforts in bringing out this useful book that would serve as reference material for various stakeholders.

New Delhi, India
June 2020

T. Mohapatra

Preface

Firstly, we convinced ourselves that there is a need for a book on "Climate-Smart Agriculture" (CSA) which is easy to understand and at the same time put the relevant information regarding challenges and opportunities of CSA in a lucid manner for researchers, students, and policymakers. The book presents the approaches and practices along with the key messages on different relevant issues on climate-smart agriculture including its barriers for adaptation and policy framework. It is primarily intended to be used as a reference book for climate change and environmental sciences. It can be used by researchers, officers, scholars, and policymakers and also could be used in undergraduate and postgraduate levels. The book should give researchers and students a good foundation in CSA and climate change studies and inspire them to take up further studies in the relevant fields.

The 12 chapters of the book systematically cover the importance, dimensions, management, barriers, mainstreaming, and policies of "climate-smart agriculture" (CSA). The chapters orient like a symphony starting from the basic concept of weather, climate, climate change, climate variability, and the cause of climate change; the impacts of climate change on agriculture, along with the needs of CSA and its dimensions; management practices in soil, water, energy, crop, livestock, and aquaculture sectors in CSA; the barriers of adaptation to CSA; and finally the policies, financing, and future needs of CSA.

In Chap. 1, the definition of climate, climate change, climate variable, and both natural and anthropogenic causes of climate change is briefed. We know that the climate change impacted negatively on food security, livelihood, and ecological balance and predicted to do so in future also. The "climate-smart agriculture" (CSA) is introduced in this chapter as a probable solution of the present-day problem of food insecurity, poor system resilience, and climate change risk. The approaches of CSA and its three pillars, namely, increase in agricultural productivity, building of system resilience, and mitigation to climate change were described with its challenges. The future need of the holistic synchronization of soil, water, energy, crop, livestock, carbon, and human resources management with planning and financing is emphasized in this chapter.

In the second chapter, the impacts of climate change, both positive and negative, on agriculture and food security are presented. The consequences of climate change, like frequent floods, drought, salinity, heat stress, tropical cyclone which threatens the food security and direct effect of temperature rise on soil carbon losses, freshwater availability, crop yield, livestock production, fish migration, and spawning, are elaborated in this chapter. Both evident and predicted impacts of climate change on agriculture and food security are discussed. The six dimensions of CSA, including water smart, crop smart, energy/carbon smart, nutrient smart, weather smart, and knowledge smart agriculture, are detailed in Chap. 3. The fourth chapter deals with the soil management for CSA which includes soil quality/health assessment, soil organic carbon management, and soil carbon sequestration. The fifth and sixth chapters elaborated the water and energy management practices under CSA. The crop management practices in CSA are the contents of chapter seven. In Chap. 7, we primarily focus on the crop managements having both mitigation potential and adaptive capacity and can play a key role in sustainable agricultural production and food security under climate change scenarios. The major such practices includes introduction of climate-stress-tolerant cultivars, development of tolerant varieties, site-specific changes in cropping sequences/ patterns, encourage the farming system approach, and propagate the ecosystem management options. Few of them are also discussed in detail, like the integrated farming system (IFS), participatory watershed management, agro-ecosystem-based agroforestry, cropping system, etc. In Chap. 8, livestock management under CSA is covered. In this chapter, different technologies and management strategies are discussed. Few of them are construction of shed and water sources in low cost to minimize the heat stress to animal, keeping high-quality breeds in lesser number, introduction of micro- and drip irrigation systems in pastures, and construction of water harvesting structures.

Chapter 9, which deals with the mitigation of climate change and GHG emissions, is the unique one that covers the supply-side and demand-side mitigation options/pathways. In supply side, the GHG emission can be reduced by land-use changes, enhancing the terrestrial carbon storage, improvement of crop and livestock management, and carbon sequestration. On the other hand, in demand-side mitigation approaches, the GHG emission can be reduced by shifting the dietary pattern (from non-vegetarian to vegetarian/flexitarian), reducing food loss/waste, and shortening the food supply chains.

The barriers to adaptation of CSA are discussed in Chap. 10. The barriers could be technology centric or user performance centric or both. The adaptation barriers to CSA are primarily socio-economic in nature both from "supply side" and "demand side." The barriers to "land-based adaptation," "policy-institutional framework," "supply-side" (technology providers) and "demand-side" (technology user) chains, and from holistic "system-functioning" perspectives are discussed in detail in this chapter. To overcome the barriers, the location-specific adaptation with "bottom-top" approach having inclusive national planning with international financing is the key to success of CSA adaptation. Mainstreaming the CSA is the key to adaptation of CSA which is presented in Chap. 11 of this book. The importance of mainstreaming of CSA in core national policy is described with its framework in this chapter. In nut shell, the policy initiatives and support of CSA have to be mainstreamed into national and regional levels having proper integration of public and private partnerships. Networking and transparent coordination among the agencies of national and local levels are the key for mainstreaming the CSA. Most importantly, the core government policies should include the planning and expenditure of CSA, so that it could fulfill the broader sustainable developmental goals of nations. Moreover, for the better adaptation of CSA, the social protection and safety measures must be an integral part of the CSA policies. In this context, we must say that the success of CSA is based on the judicious integration of social safety programs with disaster risk management strategies. In Chap. 12, the way forward message emphasizes that the CSA must follow the principles of triple win features that include gain to farmers/producers, investors, and consumers without regrets, trade-offs, and loses. Therefore, the future agriculture/generation should look for transformational changes, integration of private and public sector financing, changing the policy portfolios, and political outlooks toward CSA.

We received ungrudging help from ICAR-National Fellow Project; ICAR-NRRI, Cuttack; and a number of scientists and well-wishers for preparing this book. Those whom we wish to mention are Dr. T. Mohapatra, Dr. S. C Datta, Dr. P. Swain, Dr. M. J. Baig, Dr. D. Maity, Dr. Ch. Srinivasa Rao, Dr. D. Mandal, Dr. O. P. S. Khola, Dr. S. Kartykeyan, Dr. M. C. Manna, Mr. P. K Dash, Mr. S. R. Padhy, and Dr. S. Neogi. We have duly acknowledged the sources of the diagrams and tables that have been reproduced from other sources and publications.

Cuttack, India — Pratap Bhattacharyya
Baramati, India — Himanshu Pathak
Chandigarh, India — Sharmistha Pal

Contents

About the Authors

Dr. Pratap Bhattacharyya is currently working as ICAR-National Fellow and Principal Scientist at ICAR-National Rice Research Institute, Cuttack, Odisha, India. He obtained B.Sc. (Ag.) from BCKVV, West Bengal, and subsequently, completed M.Sc. and Ph.D. (Soil Science and Agricultural Chemistry) from IARI, New Delhi. He started his career as Scientist in 2001 at IISWC, RC, Chandigarh. His fields of specialization are climate change, climate smart rice production, soil fertility and microbiology, soil and water conservation. Dr. Pratap extensively worked on natural resource management, particularly climate change aspect on rice, climate resilient and climate smart rice cultivation in lowland rice, soil conservation, watershed management, nutrient and carbon management, and resource conservation technologies in rice. Dr. Pratap has given a unique concept that lowland rice ecology is a net carbon sink by using advanced eddy covariance techniques. He also identified the acetoclastic and serine pathways are the predominate ones in methanogenesis and methanotrophy, respectively in lowland rice ecologies by using whole genome metagenomic approach. Dr. Bhattacharyya is ICAR-National Fellow; Fellow of National Academy of Agricultural Sciences, India (FNAAS); Fellow of IASWC; got ICAR-LBS young scientist award; Dr. K. J. Tejwani award on NRM by IASWC; Mosaic Foundation young scientist award on plant nutrition; Certificate of Excellence on Eddy covariance techniques by University of Georgia, USA. Dr. Pratap has published more than 150 research articles included research papers, books, book chapters, and popular articles. He has guided 5 Ph.D. and 8 M.Sc. students. He served as associated editor for the Journal of Indian Society of Soil Science".

Dr. Himanshu Pathak is currently the Director, ICAR-National Institute of Abiotic Stress Management, Maharashtra, India. He Obtained B.Sc. (Ag) from Banaras H University, Varanasi, India; and MSc and Ph.D. in Soil Science and Agricultural Chemistry from Indian Agricultural Research Institute (IARI), New Delhi, India. His research areas include soil science, climate change and environmental pollution. He Joined as Director of ICAR-National Rice Research Institute (NRRI), Cuttack, Odisha on August 1, 2016. Earlier he served as Principal

Scientist, Indian Agricultural Research Institute (IARI), New Delhi (2009-16); Professor, Discipline of Environmental Sciences, IARI (2013-16); He also worked as a visiting scientist in University of Essex, United Kingdom; IRRI, Philippines; CSIRO Land and Water, Australia and Institute of Meteorology and Climate Research, Germany. He is Fellow of Indian National Science Academy; National Academy of Sciences, India; National Academy of Agricultural Sciences and West Bengal Academy of Science and Technology. He is the recipient of the Alexander von Humboldt Fellowship of Germany; Silver Jubilee Commemoration Medal of INSA; Golden Jubilee Commemoration Young Scientist Award of the Indian Society of Soil Science; BOYSCAST Fellowship of Department of Science and Technology (DST); Young Scientist Award, Dr. BC Deb Memorial Award, Prof. SK Mukherjee Commemoration Award and Platinum Jubilee Award Lecture by Indian Science Congress Association (ISCA) and Recognition Award of NAAS. He was the Lead Author of the Inter-Governmental Panel on Climate Change (IPCC) 5th Assessment Report. He is also the Lead Author of IPCC for updating GHG inventory. Dr. Pathak was the Coordinator for United Nations Framework Convention on Climate Change (UNFCCC), Department of Research and Education (DARE), Ministry of Agriculture and Farmers Welfare, Government of India and received IRRI, Philippines Outstanding Administrative Support Award. Currently, he is the President of Association of Rice Research Workers, Cuttack, Odisha and Executive President of Society for Conservation of Nature, New Delhi. He published more than 170 research papers, 30 review papers, 25 popular articles, 13 books, 15 technical bulletins and 75 book chapters.

Dr. Sharmistha Pal is working as Scientist in Indian Council of Agricultural Research (ICAR)- Indian Institute of Soil and Water Conservation, Research Centre, Chandigarh, India till 2008. Her fields of specialization are Soil science, soil chemistry, fertility and microbiology. She obtained B.Sc. (Agriculture) from Bidhan Chandra Krishi Viswa Vidyalaya, West Bengal, India; M.Sc and Ph.D. from Division of Soil Science and Agricultural Chemistry, Indian Agricultural Research Institute, New Delhi, India. Specifically, she worked on aluminum–organic acids interaction in rice rhizosphere in acid soil; iron dynamics under aerobic rice as influenced by its sources and methods of application; soil carbon sequestration, soil and water conservation. She got Dr. S. P. Raychaudhuri Gold Medal from Indian Society of Soil Science, New Delhi (2008); Jawaharlal Nehru Award for outstanding Postgraduate Agricultural Research from ICAR (2010); Young Scientist Award (Agriculture and Forestry Sciences) from Indian Science Congress Association (2011); Endeavour Cheung Kong Award for Post doctoral research at the University of Adelaide, Australia from Australian Government Department of Education (2014); Pran Vohra Award (Agriculture and Forestry Sciences) from Indian Science Congress Association (2016); Young Soil Conservationist Award from Indian Association of Soil and Water Conservationists, Dehradun (2017). She has published around 70 research publications including research papers, book-chapters, book, popular articles, bulletin etc.

Chapter 1
Introduction

1.1 What is Climate Change?

Weather is the short-term change in meteorological variables like temperature, rainfall, cloud, relative humidity, sunshine hours, wind velocity, etc. The weather could vary significantly even from morning to evening or from one day to the very next day. However, the climate of a city or region refers to weather averaged over three decades or more. Climate generally refers to a wider geographical boundary such as city, region, state, or country. In a matter of fact, the climate of a city, region, or continent changes very slowly and the changes take place on 30, 100, or even 1000 years' time scale. Therefore, climate change could be defined as consistent changes in the mean of the weather variables over a considerable period (minimum of 30 years). In other words, climate change refers to a long-term change in the state of the climate that could be indicated by the changes of means and/or changes in climate variability (Fig. 1.1). Climate variability refers to the variation of the climate in the mean state, standard deviation, or occurrence of extremes. Climate change also includes both abrupt and gradual changes in the intensities of extreme events.

Climate has been changing historically from ice ages to present decades of "warming-planet" due to natural as well as anthropogenic activities (human activities). However, in the last five decades, climate change is primarily due to anthropogenic emissions of greenhouse gases (GHGs).

1.2 Greenhouse Gases, Greenhouse Effect, and Global Warming

Greenhouse gases are those which can absorb and radiate back the range of infrared (IR) radiations and thereby effectively absorb as well as emit heat energy, keeping our earth atmosphere warm. Greenhouse gases in the atmosphere are having the dipole

© The Editor(s) (if applicable) and The Author(s), under exclusive license to Springer Nature Singapore Pte Ltd. 2020

P. Bhattacharyya et al., *Climate Smart Agriculture*, Green Energy and Technology,
https://doi.org/10.1007/978-981-15-9132-7_1

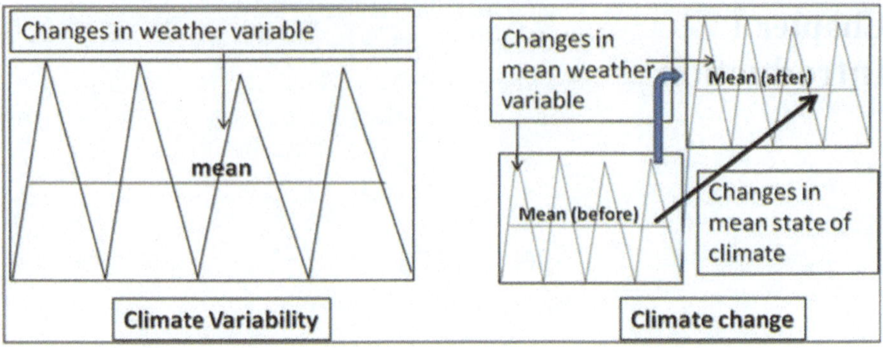

Fig. 1.1 Schematic diagrams showing the conceptual difference of climate variability and climate change

moments which make them capable to absorb and radiate back the IR radiation. Carbon dioxide (CO_2), methane (CH_4), water (H_2O) vapor, and nitrous oxide (N_2O) are the major GHGs in the atmosphere having dipole moments which are capable of absorbing IR radiations (Fig. 1.2). Although CO_2 has symmetrical chemical bonding

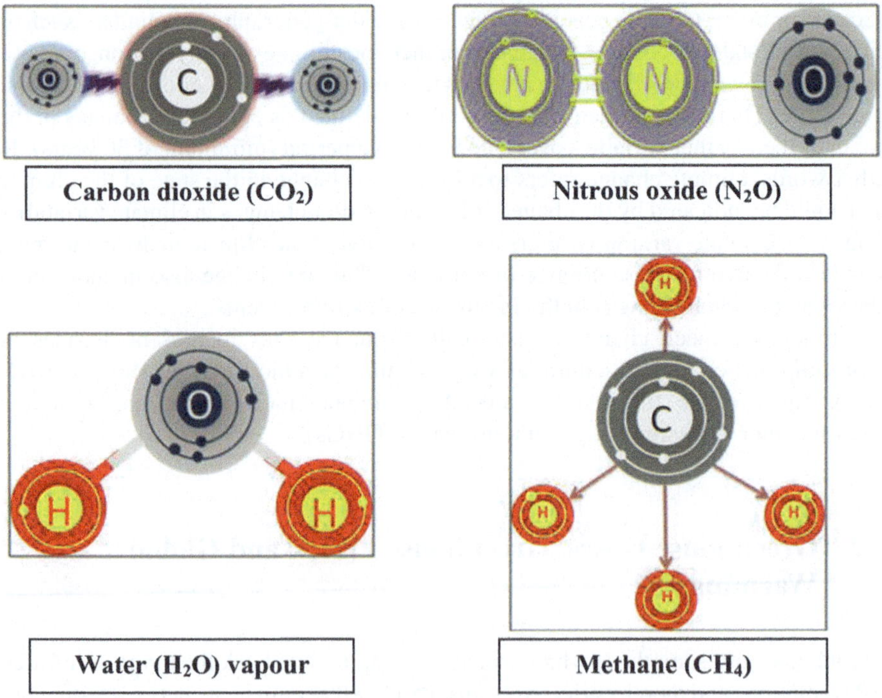

Fig. 1.2 Major greenhouse gases with the molecular structure indicating dipole moments and bonding stretch

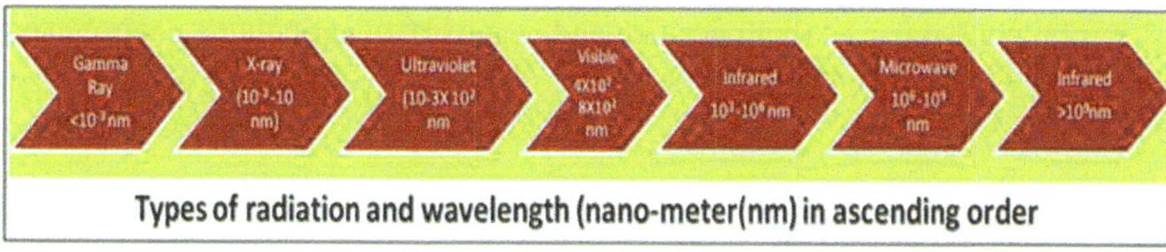

Fig. 1.3 Electromagnetic radiations with different wavelength ranges

and as such has no dipole moment, but an instantaneous dipole moment is developed due to its asymmetric vibration and stretching of bonds and this makes it a GHG. On the other hand, the two major gases in the atmosphere, namely, nitrogen (N≡N) and oxygen (O=O), were not able to absorb IR radiation as their chemical bodings are symmetrical and do not have any dipole moment.

Greenhouse gases radiate and absorb IR radiations of wavelengths in the range of 10^3–10^6 nm (Fig. 1.3). A portion of IR radiations come from the sun and others generated from the earth's surface due to the reflection of long-wave radiation. The higher energy short-wave radiations like ultraviolet (UV; 10^{-3}–10^2 nm wavelength) and visible (VIS; 4–8*10^2 nm wavelength) are absorbed by earth surface and a portion of that is radiated back to the atmosphere and during this process, those short-wave radiations lose their energy and become long-wave radiation.

1.2.1 Greenhouse Effects

In general, the term "greenhouse effect" refers to the process in which sunlight passes through the glass and warms the house covered with it. In this condition, fundamentally, the glasshouse restricts the airflow within it and thereby makes it warm. However, it is a misnomer with the original "greenhouse effect" in the earth atmosphere. The basic difference is the way a greenhouse retains heat and the process in which radiation from earth is absorbed and is radiated back by greenhouse gases in the earth atmosphere and makes our planet warm. The "greenhouse effect" is the process by which greenhouse gases in the atmosphere absorb and radiate back the long-wave radiations (preferably IR radiations) in the earth atmosphere and warms our planet surface temperature above what it might be without the presence of the atmosphere. Importantly, the surface temperature would be about 256 °K or −17 °C without any greenhouse effect and there would be no existence of life on our planet without that process. Fortunately, due to the sustained greenhouse effect on our planet, the present-day mean global surface temperature is 14.85 °C [4]. However, the mean global surface temperature has been increased by 0.85 °C in the last century due to "enhanced greenhouse effect." The "enhanced greenhouse effect" is primarily caused by anthropogenic higher emissions of GHGs which leads to an increase in global temperature. "Enhanced greenhouse effect" causes global warming which has several consequences (both positive and negative) on agriculture, environment, industry, and other sectors of our society.

1.2.2 Global Warming

"Enhanced greenhouse effect" on earth atmosphere is the resultant of increased GHG emissions due to anthropogenic activities. The "enhanced greenhouse effect" is either caused by increased well-known GHGs (water vapor, CO_2, CH_4, N_2O) present in the atmosphere or by addition of some other radiative gases which are not naturally present in atmosphere. (e.g., hydro-chlorofluro carbon (HCFC), chlorofluro carbon (CFC), etc.). The global atmospheric temperature is increasing due to this "enhanced greenhouse effect" which is called global warming. Environmentalists and climate scientists think that the human-induced greenhouse effect is the prime cause of the current trend of global warming. This thought is further strengthened as the present-day climate models could not explain the observed trends of temperature rise with the help of solar irradiation only and more so warming is only experienced at the lower atmosphere.

Very often, we use the term global warming potential (GWP) of greenhouse gases. The GWP of greenhouse gases (Table 1.1) depends on the radiative forcing of GHGs. The radiative forcing is a quantitative estimation of the effect of a particular process or factor, like land-use changes, GHGs, aerosols, etc., on the net energy balance of the earth atmosphere. It is either way, maybe positive or negative. A positive radiative forcing warms the planet surface and a negative forcing cools the surface. GHGs have a positive forcing because they work like a blanket. They restrict the direct transmission of energy into space and absorb radiating long-wave radiations from the earth's surface and warm the earth's atmosphere. The radiative forcing of GHGs (CO_2, CH_4, N_2O, water vapor, choloflurocarbon, etc.) which could absorb and radiate energy in all the directions depends on the chemical constitution of gases, their bonding strength and stretching, initial atmospheric temperature, and time scale. According to NOAA's annual GHG index, the radiative forcing of anthropogenic GHGs has increased by 27.5% from 1990 to 2009. The increased carbon dioxide

Table 1.1 Global warming potential and radiative properties of agriculturally important greenhouse gases in the atmosphere [4]

GHGs	Pre-1750 concentration (ppm)	Current tropospheric concentration (ppm)	Annual increase (%)	Atmosphere life time (years)	GWP (100 years time horizons)	Increased radiative forcing (W m^{-2})
CO_2	280.0	412	0.5	Variable	1	1.46
CH_4	0.715	1.82	0.8	12	25	0.50
N_2O	0.270	0.325	1.0	114	298	0.15

Source IPCC [4]; NOAA-ESRL [5]

generated by fossil fuel burning and energy sector is responsible for 80% of that enhancement. Whereas, the contribution of methane and CFCs to radiative forcing has been more or less the same in the last three decades. On the other hand, aerosols or small particles have both positive and negative radiative forcing. Those properties depend on their pattern and mode by which they reflect light or absorb and emit heat. We know that the black carbon aerosols absorb sunlight and have a positive forcing. On the other hand, the sulfate aerosols directly reflect sunlight back into space and have a negative forcing.

1.3 Causes of Climate Change

Climate has been changing since the inception of the earth. Earth's temperature and precipitations have been influenced by continental drifts and volcanic eruptions, which released a large volume of CO_2 which consequently heated the earth's atmosphere. Today also, these natural phenomena make a significant influence on the climate. For example, El Niño occurs at intervals of 3–7 years, when the trade winds ease, and the warm water from the Western Pacific moves toward east resulting in a rise in sea temperature in the west of Peru. The El Niño also leads to significant deviations in precipitation and temperature worldwide. Therefore, causes of climate change are wide and different and at the same time, the impacts of that on our various ecosystems are complex.

If we bulleted the three basic factors of climate change they are, namely, the variations in the sun's energy received by the earth, fluctuations of reflectivity of land surface and earth atmosphere, and changes in the greenhouse effect. The direct measures of climate change are the carbon dating of biological material and the previous records of earth's climate data. However, scientists have used several indirect measures to detect climate change like pollen remains, ocean sedimentation, changes the earth's orbit angles around the sun, ice cores, glacier lengths, tree rings, etc.

If we critically analyze, then we can see the climate changes before the industrial revolution (1750) primarily because of natural causes (Fig. 1.4). However, recent climate changes, i.e., during the 1900s, and after, the phenomenon cannot be explained only through natural cause rather than it suggested that anthropogenic activities are primarily responsible for that. It is a reality that humans' activities (anthropogenic) have been impacting the climate since the industrial revolution (1750). The mean global temperature has risen by 0.85 °C since the industrial revolution [4] and the sea level has risen by around 20 cm. More specifically, we can say up to 1950 the influence of nature was more significant than human activities. However, after that, the trend of climate change only could be explained precisely by sector-wise anthropogenic activities. According to the 5th IPCC report, it is more than likely (>90%) that most global warming and climatic changes are attributed to increasing anthropogenic GHGs (Fig. 1.4). However, the magnitude of global warming in the future is uncertain. Primarily, still we do not have a clear-cut idea about, how much GHGs would be going to increase in the future (depending on

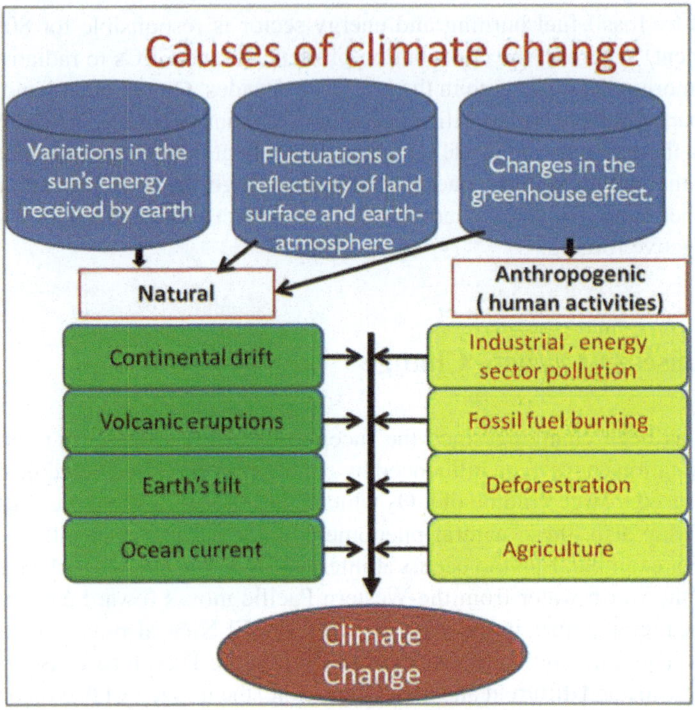

Fig. 1.4 Causes of climate change

technological advancement and economic growth), and secondly, we do not know the climate sensitivity and/or buffering capacity of our complex climate systems.

1.3.1 Changes in Solar Energy and Tilting of Earth Orbit

Changes in solar energy intensity are one of the major causes of natural climate change. Changes in the sun itself lead to the variation of emitted radiations and its energy- intensity outputs. Warming of earth's atmosphere usually occurs during the period of stronger solar intensity and cooling at the time of weaker solar intensity. In general, there is a natural 11-year cycle of ups and downs in solar intensity. Tilt position, the shape of earth axis, and orbits influenced the amount of sunlight/energy which reached the earth surface and have their impacts on natural climate change.

In the past, "little ice age" (seventeenth–nineteenth centuries), there was a cooling effect of solar intensity which caused world climate cooler, particularly in Europe and North America. In true sense, the long period of cooling during ice ages and shorter warming period in the intermediate ice age are basically due to tilting of earth orbit and changes in solar intensity. Both cases have been considered as a natural cause of climate change. As the change of solar intensity and tilting of earth orbit are very long-term phenomena, therefore, to judge the effect of those on climate changes in a smaller time scale in the recent past is not proper and also not significant.

1.3.2 Changes in Reflectivity Earth-Atmospheric System

Some portions of the sunlight are absorbed by the earth's surface and some are reflected back to the atmosphere. The absorption and reflection capacity of surface and earth components varies with their color, surface roughness, and chemical makeup. Light-colored materials and surfaces like white snow and clouds reflect a higher amount of light, whereas dark-colored objects like soil, forest, and ocean absorb the majority of sunlight falls on it. Albedo, which is the ratio of reflected and incident radiation of an object, is a measure of reflection property of the object. As a whole, earth albedo is about 30%. Not only land and water, but aerosol also plays an important role in sunlight absorption or reflection and thereby impacts climate change. For example, aerosol formed by sulfur emission from coal burning and particles from the volcanic eruption has a cooling effect on earth's atmosphere as they reflected back much radiation from the sun to space.

In the past, natural changes are the reflectivity of surface/objects governed by the melting of sea ice and aerosol formed by particles erupted from volcanoes, which have both warming and cooling effect but very slow in the process. However, in the recent past, changes in land use, land cover, deforestation, desertification, urbanization, and human-induced generation of aerosol are responsible for climate change. These changes are relatively rapid. Sometimes, human-induced aerosols cool the earth's atmosphere also and counteract the warming processes. Therefore, the net feedback mechanism and its impacts are getting more importance nowadays rather than a single component effect for explaining the cause of climate change.

1.3.3 Changes in the Greenhouse Effect

In the past millions of years, the atmospheric carbon dioxide levels on our planet primarily vary with changes in glacial cycles. The CO_2 levels were higher in the "interglacial" period and lower in the "cool glacial" period. The heating and cooling of the ocean and land surface also influence the natural concentrations of GHGs in the earth's atmosphere. However, since the "Industrial Revolution" (1750) the anthropogenic activities (human activities) contributed significantly to GHG emissions which lead to global warming and climate change. The atmospheric CO_2 concentrations have increased by more than 40% since pre-industrial times, from approximately 280 parts per million by volume (ppmv) in the eighteenth century to over 412 ppmv in 2019 [5]. The current CO_2 level is higher than what it has been in at least 800,000 years. The anthropogenic activities like fossil fuel burning, and industrial pollution, power sector emission, deforestation, and land-use change resulted in the "enhanced greenhouse effects" and climate change. The U. S. Geographical Survey (USGS) [6, 8] reveals that human activities now resulted in 135 times more CO_2 emission (30 billion tons per year) per year than caused by volcanic eruptions each year. Methane is produced naturally from wetlands and anthropogenically by agricultural activities, rice-paddy cultivation, and fossil fuel extraction. Like CO_2, present CH_4 concentration in the atmosphere is also measured highest since the past 80, 0000 years. However, during the twentieth century, its concentration has been grown maximum which is around 2.5 times to that of the pre-industrial period. But fortunately, in the last three decades, the rate of increment of CH_4 concentration in the earth's atmosphere has slowed down. Naturally, nitrous oxide (N_2O) is produced by biological processes. Other significant human-induced processes are fossil fuel burning and indiscriminate use of N-fertilizers. Similar to carbon dioxide and methane its concentration has also increased by 20% since 1750 (Industrial revolution).

Among the other greenhouse gases (GHGs), water vapor, the most abundant natural GHG has a natural greenhouse effect. By and large, global water vapor concentration is not significantly affected by human activities. On the other hand, tropospheric ozone (O_3) generated by chemical reaction of nitrogen oxides and volatile organic compounds (generated by automobile, power plant, and industry) with sunlight causes climate change more frequently in recent past [3, 4, 8, 6].

1.3.4 Feedback Mechanism as a Cause of Climate Change

The majority of climate scientists in the present day believe that the net feedback effect of different factors of climate change is more valid than a single factor. Climate change feedback could amplify or reduce the warming or cooling effect caused by the single factor of climate change. If the feedback facilitates climate change, it is called positive feedback. And if retarded the changes, it is termed as negative feedback. Therefore, the feedback mechanism is taken into consideration of all significant causes/factors of climate change like solar intensity, reflectivity, aerosols behavior, GHG concentration, carbon cycle, etc., all together at a particular time. As for example, although evaporation has a cooling effect, water vapor (being a GHG) enhances global warming and provides positive feedback. On the other hand, some type of highly reflective cloud causes negative feedback to warming (reflected more sunlight to space). Therefore, permafrost thawing, ocean absorption of CO_2, S-particle-produced aerosol, relative GHG concentration as well biotic and microorganisms diversification in the soil must be considered when judging the processes for positive and negative feedbacks to climate change.

1.4 Impact of Climate Change on Agriculture

The impact of climate change on agriculture is frequently described as consequences of climatic extremes on agriculture. On many occasions, it is also viewed as changes in weather variables and their subsequent effect on crops and livestock. Changes of weather variables due to climate change could be listed as increase in mean air temperature, sea-level rise, changes in precipitation patterns, increased fluctuations of both in atmospheric air temperature and rainfall events, lowering of the groundwater table and drying up of water bodies, increase in frequency and intensities of extreme events such as droughts, floods, heat waves, cyclone, etc. Further, impacts of drought and floods also could be characterized as reduced availability of food, water and nutrition, destabilization of the economy, and food insecurity.

Broadly, we can say the crop yield would be reduced in currently warm and low rainfall areas. However, presently low rainfall regions having higher projection in the future are likely to be benefited. Similarly, regions with sub-optimal temperatures would become more favorable for plant growth and yield. The C3 crops (e.g., rice, wheat) may benefit more than the C4 crops (e.g., maize, sugarcane) due to the CO_2 fertilization effect under elevated CO_2 conditions. Recent studies on tropical rice system of eastern India revealed that there was an increase of grain yield by 22.6% when exposed to elevated CO_2 (550 ppm) but no detrimental effect on the reduction in grain yield under elevated temperature (2°C more than ambient) in wet (Kharif) season [7]. However, crops may require additional fertilizers to reap the benefits. Therefore, GHG-induced changes are not just about negative impacts but also may bring some positives as well. Challenge is to cope with the sequential and multiple stresses on crops which are causing significant loss to crop production. Adaptation strategies should address these challenges.

The effects of elevated atmospheric CO_2 concentration on wheat physiology and yield were studied by the meta-analysis of datasets. The results showed that elevated carbon dioxide (CO_2:800 ppm) significantly increased wheat grain yield by 24% with 95% confidence levels [9]. Many enclosure studies also indicated an increase of maize growth under elevated CO_2 and well-watered conditions, while others reported no significant effects of elevated CO_2. In "free-air-carbon dioxide enrichment" (FACE) experiments with C4 crops, only biomass of sorghum was slightly increased by CO_2 enrichment; however, biomass and grain yield of maize were totally unaffected under sufficient water availability. Under drought stress, the growth of maize was generally increased by CO_2 enrichment and the relative CO_2 effect was greater under well-watered conditions. Further, GHG-induced climate change is projected to alter the quality of grain. Changes in the biochemical composition of tissues may alter the crop-pest interaction in future climates. On the other hand, the plantation crops and horticultural crops are also influenced in terms of phenological events, yield, and quality of produce. Since plantation crops are perennial in nature and live through climatic stresses and climate change vagaries, the adaptation options need to be multidimensional.

Impacts of climate change on livestock are variable. Increase of temperature and humidity stress on buffaloes and cows likely to reduce the milk yield. The marine and inland fishes are affected in terms of change in the breeding season, increase in the growth rates and breeding cycles, and horizontal and vertical extension of fish habitats. Poultry birds may have to face more heat stress in the future making it necessary to develop efficient sheds.

We have seen in the previous two sections that climate change is real and which is also changing at a rapid rate affecting all forms of agriculture both in positive and negative directions. However, negative impacts are more alarming which need quick and sustainable solutions. Moreover, solutions also should not further induce climate change. Hence, climate-smart agriculture is one of the answers and also need of the day which is, on one hand, climate-resilient and, at the same time, addresses the issues of food security and GHG mitigations. In the coming sections, we will briefly discuss the necessity of climate-smart agriculture, its relevance, and dimensions.

1.5 Why Climate-Smart Agriculture

Now the question is why climate-smart agriculture? And what is the need for CSA in the present day? If we see critically, the world population is expected to increase by 1.3 times by 2050 (from 7.39 billion (2018) to 9.7 billion (2050)) and at the same time, agricultural production will have to increase by at least 60%. Further, climate change is a threat to food and nutritional security under a "business-as-usual" approach, and therefore adaptation, mitigation, and reduction of GHG emissions are necessary. In this context, the CSA approach would open up the window for increased food productivity, sustainability, resilience, and food security of the system, simultaneously.

To understand the approach to be followed in CSA, we must know the history of CSA. The term climate-smart agriculture (CSA) was coined in 2009. Subsequently, the concept of CSA was presented in the 1st global conference on "Food Security, Agriculture, and Climate Change" in Hague 2010. In 2012, at the 2nd global conference in Hanoi, Vietnam, the sourcebook for CSA adopted the concept to benefit primarily the small and marginal farmers and vulnerable people in developing countries. In the very next year in 2013, at 3rd global conference in Johannesburg, South Africa, the effort and discussion were initiated regarding the alliance of CSA. Finally, in 2014 at Climate Summit, in New York, the global alliance for the CSA action plan was presented.

1.6 Definition, Pillars, and Dimensions of CSA

1.6.1 Definition of CSA

Climate-smart agriculture could be defined as "an integrated approach for developing technical, policy, and investment conditions to achieve sustainable agricultural development for food security under climate change" [2]. It is a relatively new approach to develop the technical, political, and financial conditions for the achievement of sustainable development goals [1]. There are three pillars of CSA, namely, (i) sustainably increasing agricultural productivity and incomes, (ii) adapting and building resilience to climate change, and (iii) reducing and/or removing carbon emissions.

Precisely, FAO has forged the concept of climate-smart agriculture (CSA) as a way forward for food security in a changing climate. The "CSA aims to improve food security, helps communities to adapt climate change, and contributes to climate change mitigation by adopting appropriate practices, developing policies and mobilizing needed finances" [2].

1.6.2 Dimensions of Climate-Smart Agriculture

The key dimensions of CSA refer to the system which is water smart, weather smart, nutrient smart, energy smart, carbon smart, and also knowledge smart. The important water-smart technologies/practices in agriculture are direct-seeded rice, raised bed for maize, the system of rice intensification (SRI), and drip irrigation particularly for fruits and vegetable crops. Weather smart CSA refers to robust weather forecasting and imbedded agro-advisory services, index or real-time crop/livestock insurance scheme, and contingency seed supply facilities. There are various techniques available for making CSA really nutrient smart, those are real-time nitrogen management through leaf color chart, SPAD meter-based nitrogen supply, sensor-based major and micronutrient application in precision farming, etc. Carbon and energy-smart agriculture indicates the practice of residue management and the adoption of zero tillage. On the other hand, knowledge smart refers to practices of Information and Communication Technologies (ICTs), gender equality, and capacity building through training and awareness campaigns. Nowadays mobile apps for pest and disease management gained popularity among different stakeholders and farmers. Social and gender issues are important in CSA, as agriculture women workers are the backbone of the rural economy, and women farmers represent more than a quarter of the world's population. More so, women comprise around 43–45% of the agriculture workforce in developing countries. However, they have less access than men in agriculture-related assets, inputs, and services. Therefore, as recommended by FAO, to make CSA gender smart, policy recommendations are obligatory for gender analysis within all CSA projects and programs, and to identify women groups and provide them adequate support and proper need-based training for negotiation, leadership development as well as business and communication skill. Few other issues are, support women to participate in the decision-making process related to climate change at the local level and to facilitate women's access to credit and land through amending laws.

1.7 Brief Nodal Approaches of CSA

Climate-smart agriculture is an integrated approach to synchronize three dimensions of sustainable development, namely, economic, social, and environmental by addressing the issues of both food security and climate challenges. It should bring together all the three aspects of management, like practices, policies, and institutions. Therefore, all the components of agriculture, viz, soil, water, energy, crop, and livestock must be managed judiciously with balanced synchronization in CSA (Fig. 1.5).

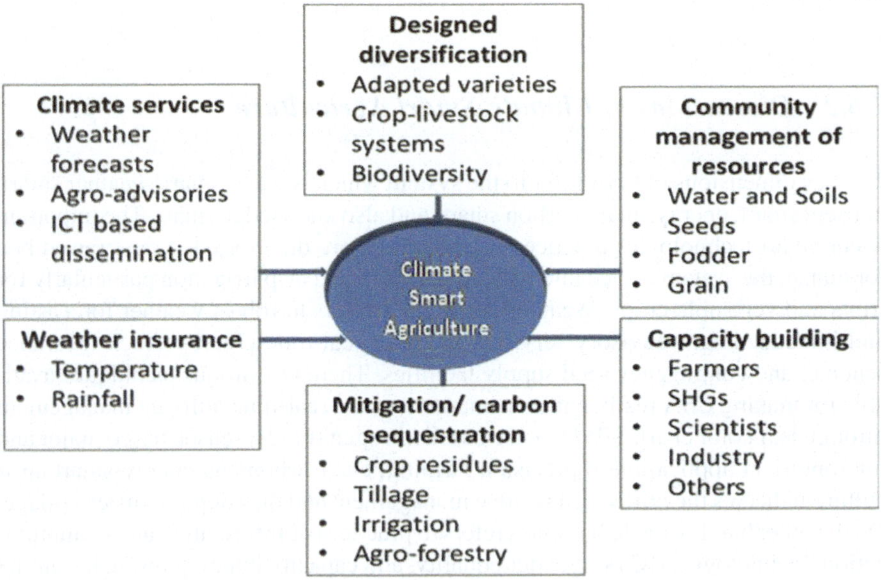

Fig. 1.5 Different components and options/practices of CSA

1.7.1 Soil and Water Management

Soil and water management have various facets. If we are talking about system-based management then conservation agriculture (CA) is a well-proven option. Conservation agriculture has three integrated components, viz, minimum soil disturbance (zero tillage, minimum tillage), residue retention, and crop diversification. This system not only saves energy but also sustains soil and environmental quality and promotes carbon sequestration. However, the increase in productivity of the system is still an issue in CA. If we talk about the component-wise soil and water management, then rainwater harvesting, contour bunding, contour planting, alternate wetting and drying, SRI method of cultivation, and micro-irrigation (drips and sprinkler) could be listed on a region-specific demand and suitability. However, each component is having its own inbuilt limitations. In a soil conservation point of view, integrated watershed management is an excellent option. It takes care of local people interest, small and marginal farmer's livelihood, check soil and water erosion, increases water productivity, catchment plantation (environment protection and maintenance of biodiversity), reduction of greenhouse gases (GHGs) emission and economics of the watershed.

Assessing and characterizing (knowing) the status of soil properties is fundamental for making decisions about sustainable soil management practices. Soils that have been degraded are more vulnerable to the negative impacts of climate change. The depletion of soil organic matter (SOM) and biodiversity, greater soil compaction, and erosion are the major issues there. In addition to that, land degradation is itself a major cause of climate change. Practices/management increases SOM (precisely soil organic carbon (SOC)) which is the key to sustaining ecosystem functions. Integrated soil management practices could create optimal physical, chemical, and biological conditions for sustaining agricultural production and maintaining soil health.

The design of CSA strategies must be viewed through "water lenses." Both *rainfed* and irrigated agriculture have been affected by climate change through increased crop evapotranspiration, erratic precipitations including rainfall, variations in river runoff, and groundwater recharge. Water-smart CSA also includes a range of response options related to technologies, investments, policies, and sector-wise water management (institutional, industrial, agriculture). Further, those options would need to be applied at different scales: on local farms, in small irrigation schemes, in watersheds, in regional scale (river basins), and at the national level.

1.7.2 Crop Management

Crop management in CSA primarily focused on profitability, stress (abiotic and biotic) tolerance variety development and adaptation, crop rotations, intercropping with legumes, germplasm conservation, and improved storage and processing techniques for minimizing the losses. The selection of climate-resilient and stress-tolerant varieties is the primary management strategy of CSA. As, for example, rice cultivars like "Swarna Sub-1" for submergence tolerance, "Sahabhagidhan" for drought tolerance, "CR Dhan 801" and "CR Dhan 802" for both submergence and drought tolerance, etc. Introduction of legumes (green gram, black gram, lentil, etc.) in cereal-cropping sequences not only increases the productivity and resilience of the system, but also maintains soil fertility and reduces GHG emissions (in terms of CH_4 in the rice-legume system as compared to the rice-rice system in eastern India). More importantly, cropping systems to be followed at the specific zones in the country need to be done based on climate, water availability, and soil characteristics to make our agriculture climate smart.

1.7.3 Livestock Management

Livestock is an integral component of agriculture. It not only provides livelihood security but also maintains ecological balance. However, we should remember that ruminants are contributing around 53% of total CH_4 emission from Indian agriculture. Some of the management options are

(i) Improved feeding strategies (like, cut and carry) of cattle;
(ii) Introduction of rotational grazing both in watershed and pastures;
(iii) Improvement of the quality and quantity of fodder crops and make it part of the main cropping sequence;
(iv) Conservation and restoration of grassland; and
(v) Improvement of the shelter of livestock and their health.

In this connection, it is important to mention that the management of agroforestry and social forestry is necessary for CSA. This provides component supports for both crop and animal husbandry. Few options related to agroforestry management can be listed as boundary and hedgerow planting, planting and preservation of nitrogen-fixing trees in farmlands, the introduction of multipurpose trees in watershed/farmland, and development of woodlots and establishment of fruit orchards in classified land (e.g., land capability classes: IV, V, VI).

References

1. Food and Agriculture Organization (FAO) of the United Nations (2015) AQUASTATDFO. AQUASTAT, (Database). https://data.fao.org/ref/75f7d9c5-57ab-4a62-88d3-1e47fb50c45
2. Food and Agriculture Organization (FAO) of the United Nations (2013) Climate-smart agriculture-sourcebook. E-ISBN978-92-5-107721-4
3. IPCC (2013) Climate change 2013: the physical science basis. Contribution of working group I to the fifth assessment report of the intergovernmental panel on climate change. In: Stocker TF, Qin D, Plattner, GK, Tignor M, Allen SK,. Boschung J, Nauels A, Xia Y, Bex V, Midgley PM (Eds). Cambridge University Press, Cambridge, UK and New York, NY, USA

4. IPCC (2014) Summary for policymakers. Climate change 2014: Impacts, adaptation, and vulnerability. Part A: Global and sectoral aspects. In: Contribution of working group IIto the fifth assessment report of the intergovernmental panel on climate change. Cambridge University Press, Cambridge, UK and New York, NY, USA, pp 1–32
5. NOAA-ESRL (2019) https://www.co2.earth/daily-co2
6. NRC (2010) Advancing the science of climate changes. National Research Council. The National Academies Press, Washington, DC, USA
7. Roy KS, Bhattacharyya P, Neogi S, Rao KS, Adhya TK (2012) Combined effect of elevated CO_2 and temperature on dry matter production, net assimilation rate, C and N allocations in tropical rice (*Oryza sativa* L.). Field Crops Res 139:71–79
8. USGCRP (2014) Climate change impacts in the united states: the third national climate assessment. In: Melillo JM, Richmond TC, Yohe GW (Eds) U.S. Global Change Research Program
9. Wang L, Feng Z, Schjoerring JK (2013) Effects of elevated atmospheric CO_2 on physiology and yield of wheat (*Triticum aestivum* L.): a meta-analytic test of current hypotheses. Agric Ecosyst Environ 178(5)

Chapter 2
Impact of Climate Change on Agriculture: Evidence and Predictions

2.1 Introduction

Climate change affects agriculture and food security both negatively and positively. Direct consequences of climate change such as increases in drought frequency, extreme weather events (cyclone, heat waves, etc.), the intensity of rainfall, frequent dry spells, flash floods, sea-level rise, and sudden thawing of permafrost could trigger the land degradation processes. Anthropogenic land degradation is prevalent in the majority of ice-free land areas [11, 33]. The coastal region and drylands are more vulnerable to erosion and human-induced land degradation. During 1961–2013, the frequency of drought has increased by ~1% in drylands globally and around 380–620 million people (during the period between the 1980s and 2000s) have been affected by desertification in those areas. The most vulnerable regions for desertification were the North and Sahara region of Africa, the Middle East, the Arabian Peninsula, and Southeast Asia.

Specifically, the mean global air temperature on land surface has increased by 1.53 °C (±0.15 °C), while world's surface (including ocean and land) air temperature has increased by 0.87 °C (±0.12 °C) [65]. As a whole, the surface air temperature (land only) has increased by two-fold in recent years as compared to the pre-industrial era (1750) caused global warming at an alarming rate in the last century. Subsequently, global warming exacerbated the frequency of high-intensity rains, drought, dry spell, heat waves, and related damage to agriculture and food security. The pieces of evidence and predictions of direct and indirect impacts of climate change on agriculture and food security are elaborated in this chapter with examples.

© The Editor(s) (if applicable) and The Author(s), under exclusive license to Springer Nature Singapore Pte Ltd. 2020

P. Bhattacharyya et al., *Climate Smart Agriculture*, Green Energy and Technology,
https://doi.org/10.1007/978-981-15-9132-7_2

2.2 Evidence and Predictions of Climate Change

2.2.1 *Pieces of Evidence*

The global picture showed that during 1961–2003, the drought events have increased in Africa, west and northeastern Asia, South America, and the Mediterranean region. As a consequence, desertification has increased in parts of east and central Asia and Sub-Saharan Africa during the last three to four decades. On a global scale, if we see the trend, the contraction of polar-climate zones and the expansion of arid-climate zones are evident due to global warming in the last four to five decades. Therefore, shifts of seasonal activities and changes in abundances of flora and fauna of specific regions were observed. The occurrences of heavy intensity rainfall have also been increased on a global scale. Satellite imageries revealed the "*Vegetation-Greening* (VG)" in Europe, South America, central of North America, Asia, and southeast Australia due to extended growing season, increased nitrogen deposition/fixation, and carbon dioxide fertilization. Whereas "*Vegetation-Browning* (VB)" was noticed in northern Eurasia, Central Asia, and a few parts of North America, primarily due to water stress. However, global warming is a predominant trend but fortunately remote-sensing-based satellite imageries showed a relatively larger patch of "*Vegetation-Greening* (VG)" compared to "*Vegetation-Browning* (VB)" globally in last three decades. The alarming effect of sea-level rise has been causing higher coastal erosion and putting unprecedented land-use pressure on coastal regions of South America, Asia, Australia, and Africa.

2.2.2 *Predictions*

The majority of climate change simulation models and scenario analysis predicted that climate change would be impacted negatively on land, biodiversity, livelihoods, and human on some specific vulnerable regions of the world in the twenty-first century and exerted higher stresses on natural resources. For example, the frequency as well as the intensity of droughts is projected to increase in the Mediterranean region and southern part of Africa. The enhanced anthropogenic greenhouse gas (GHG) emissions (GHGs like CO_2, CH_4, N_2O) would enhance global warming and subsequently triggers the desertification leading to the reduction of vegetation cover in certain regions. At the same time, in a trade-off mechanism, this decrease in vegetation cover would tend to increase the local *albedo* that may lead to surface cooling. The projected thawing of permafrost due to global temperature rise is expected to increase the loss of soil carbon (primarily organic soil carbon) reserve. On the contrary, during the twenty-first century, higher vegetation growth in those regions (permafrost-thawed region) may compensate for the loss of soil carbon marginally. Specifically, at the current levels of global warming (~1.5 °C), the projected risks of permafrost degradation, water scarcity in dry land, and food supply instabilities are high. However, warming at ~2 °C, the risk from permafrost degradation and food supply instabilities is projected to be very high. Moreover, the risk of vegetation loss, wildfire damage, and dry land water scarcity is projected to be extremely high at ~3 °C global warming [40, 41, 9]. Apart from agriculture, urbanization and global warming would heat up the cities and their surroundings. As per models' predictions, the nighttime temperatures would be affected more than that of daytime temperatures that would make the city life uncomfortable.

2.3 Impact of Climate Change on Crop

Climate change impacted both negatively and positively to field crops, horticultural crops, pastures, and forestry. In this chapter, we would primarily discuss the impact of climate change on major field crops like maize, wheat, rice, soybean, potato, few major horticultural crops, etc.

2.3.1 Pieces of Evidence

Overall, the yields of maize and wheat decreased at the lower latitude globally; however, in few areas, the yields of maize, wheat, and sugar beets have enhanced at higher latitude in the last three to four decades. From 1981 to 2010, the global mean yields of wheat, maize, and soybeans decreased by 1.8, 4.1, and 4.5%, respectively, as compared to pre-industrial time even considering the CO_2 fertilization and agronomic adjustments. The yield impacts were −7.5 to + 4.3% for wheat, −8.5 to + 0.5% for maize, and −8.4 to −0.5% for soybeans (Fig. 2.1). However, no significant impacts

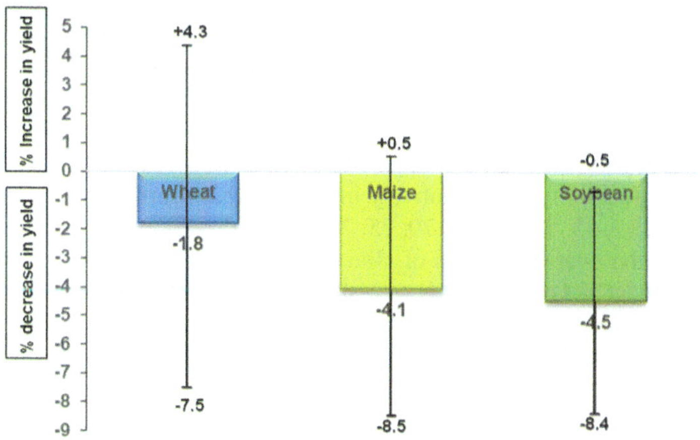

Fig. 2.1 Climate change impacts on three major crops

on the mean yield of rice were reported. This impact assessment study revealed that even the technological boost-up of "genetical-yield" of the crop in recent times and adaptation (at present rate) could not able to offset the negative impacts of climate change on yields in lower latitudes.

In this context, discussion on some specific case studies throughout the world would be useful. In New South Wales of Australia, low rainfall in combinations with high temperature at the reproductive stage of wheat has decreased yields, and high temperatures were found most detrimental [39]. Similarly, the average wheat yields in India have reduced by 5.2% from 1981 to 2009 due to warming (both daytime and nighttime temperatures have risen) and even some adaptation measures were undertaken [27, 43]. The rice areas were expanding in the northeastern and northern regions of China mostly after the year 2000 [48, 78, 80]. The yield of rice was also higher in those regions over a similar period [78]. However, the other school of thoughts stated that not only regional warming (1.43 °C in the last century) but also the other factors, like technological progress, structural changes, and government policies [24], have contributed to the observed expanded rice areas and yield in those regions. The crop phenology of winter wheat was changed [31] in northwest China from 1981 to 2011 in the aspect of delayed emergence, advanced anthesis, maturity, and prolonged reproductive stages [47]. The phenology and productivity of spring cotton were also changed in northwest China during that period [34]. However, in contrast, few reports showed a positive impact on wheat production in northern China and negative in southern China from 1981 to 2009, [74]. In a similar node, an increase and variability in soybean and maize yield were reported in South America (Argentina) during those periods [37] which have significant implications on agriculture, biodiversity, and human health [68]. However, the point of concern is that in Africa, the yields of wheat, maize, and sorghum have decreased consistently in recent years due to climate change, which is further widening the gap of food insecurity there [44]. Specifically, the negative impact of climate change on livelihoods was visible to arable crop farmers in Nigeria, Africa [1, 36, 57]. At the same time, increasing incidence of malnutrition was noticed in Cameroon, Africa which was partly due to the negative influence of extreme drought on agriculture [12].

The climate change-vulnerable mountain regions of India, Nepal, Pakistan, and China (Hindu-Kush-Himalayan region) have been experiencing an increase in climatic extremes. The people and the farmers of those regions have been facing prolonged droughts and higher frequencies of flash floods. As a consequence, the agricultural productivity was negatively impacted which further increases the food insecurity [20, 17, 35, 51, 58, 59, 70, 75, 79].

In Europe, the overall impacts of climate change on agriculture varied [53]. In general, the wheat and barley yields were decreased by 2.5% and 3.8%, respectively, and the yield of maize and sugar beet was increased slightly since 1989. However, the impacts were small and not evenly distributed. The warmer southern Europe has suffered more than the cooler northern Europe due to global warming. Even in cooler regions like the United Kingdom and Ireland, the negative effect of warming was somewhat nullified with the increase in rainfall in those regions.

2.3.2 Predictions

The future impacts of climate change on agriculture throughout the world have been studied by the approaches like, "global gridded crop model simulations [16]," "temperature-yield regression models [3]," "point-based crop model simulations [2]," and through "point-based observations in the field [82]." Irrespective of the methods/approaches followed the negative temperature impacts on crop yield were reported consistently on a global scale [81]. Similar kinds of impacts were also predicted in the city and country level. However, yield responses are not always linear but vary with growing season and temperature in specific sites. Overall, the projected global mean yield of soybean and maize would decrease at the end of this century due to warming, while the productivity of rice and wheat would increase (up to 3 °C rise in temperature in 2091–2100 as compared to 1850–1900, [38]). The projected yield would be more negative in tropics as compared to mid and high latitude and more pronounced in nitrogen-stress situations [66, 46]. In different temperature change scenarios, like, 1, 1.5, 2, and 4 °C, warming the region of northern Africa and Middle East would suffer from yield decrease with increasing temperature regardless of adaptation measures taken and or CO_2 fertilization [61, 77], while, in Latin America, the crop yield would change positively with CO_2 fertilization and negatively with temperature rise [60]. Further, the model-based Representative Concentration Pathways (RCPs) scenario prediction revealed that there would be little effects on peanut yield in Senegal, West Africa due to CO_2 fertilization and/or temperature rise [22]. However, the special report of IPCC on global warming stated that, overall, the climate-related risks to food security are projected to increase with the warming of 1.5 °C and would increase further with 2 °C in the coming century [40].

Agriculture of the Indian subcontinent is highly vulnerable to climate change vagaries as 80% of farmers of the country are either small or marginal (<2 ha landholding), heterogeneous, unorganized, and having poor coping capacity. The majority of the area is *rainfed* (~60%) frequently affected by the flood, submergence, salinity, drought, cyclones, heat waves, hailstorm, etc. In general, an increase in surface temperature could alter photosynthesis efficiency, reduce crop duration, and increase plant respiration rates. Further, an increase in air-soil-plant canopy temperature would also increase evapotranspiration and trigger nutrient mineralization in soils which may lead to lower nutrient use efficiencies. Higher frequency floods, intermittent droughts, soil erosion, and poor quality irrigation water as indirect consequences of climate change would negatively impact the crop productivity in India (Fig. 2.2). However, the CO_2 fertilization could increase the yield of rice and wheat (C_3 plants) up to certain degree rise in temperature in coming decades, but despite this fertilization effect the wheat yield is likely to be reduced due to decrease in crop growth duration, increased photorespiration, and water-stress situations in Indo-Gangetic Plain (IGP) of India. *Rainfed* crop production (around 60% area) in India would be more vulnerable due to delayed monsoon, changing rainfall pattern, terminal drought, and alternate flood-drought condition during the cropping

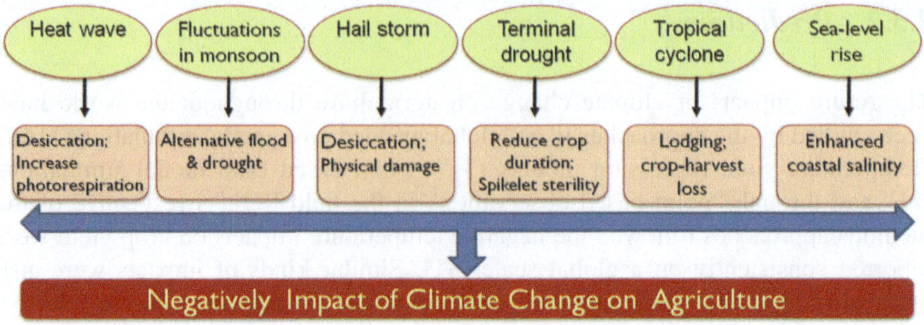

Fig. 2.2 Climatic aberration and potential negative impact of climate change on crops

season. Enhanced coastal salinity in the parts of Odisha, West Bengal, Tamil Nadu, Andhra Pradesh, and Andaman and Nicobar Islands due to intrusion of tide-water and sea-level rise would negatively affect the crop yield in those regions.

Horticultural production is also badly affected by climate change issues and consequences. Specifically, the quality of coffee, tea, fruits, vegetables, medicinal plants, and aromatics is already showing decreasing trends in many parts of India due to global warming and erratic seasonal behavior. However, in contrast, the yield loss of mustard and vegetables due to frost damage and cold waves (less frequency) could be decreased in northern India. The rapid development of fruit set during fruiting of annual vegetables and fruits due to shortening the time of photoassimilation because of unusual heat stress impaired the product quality and reduced the yield. Projection tells that vegetables growing in the relatively cooler region (mean average temperature, ≤ 20 °C) would be affected less (only 5% or so) due to global warming of 4 °C. However, yield loss of vegetables can go up to 31.5% in hotter baseline temperature (>20 °C) due to the same warming scenario (4 °C) in the coming century [69]. Another long-term predictions (2078–98) showed that there would be 20–40% and 30–60% loss of viability in common bean- and banana-growing areas in Africa, due to global temperature rise of 2.6 °C and 4 °C, respectively [62]. Moreover, fruits and vegetable production is likely to be critically vulnerable to climate change at their reproductive stages. Predictions show the same trend and also pointed out the potential threats for higher pest pressure [76].

2.4 Impacts of Climate Change on Livestock, Aquaculture, and Fisheries

2.4.1 Pieces of Evidence

The four key factors for livestock productions, namely, animal growth, animal reproduction, animal health, and water availability, are affected by temperature and heat stress (Fig. 2.3). Livestock productivity directly depends on the production and

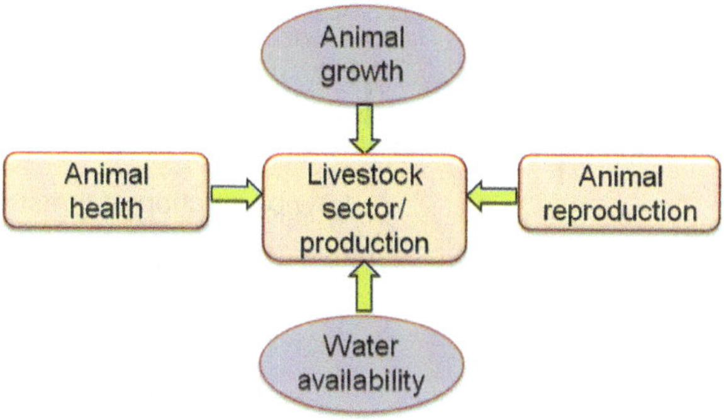

Fig. 2.3 The four key factors of livestock production

quality of pastures and rangelands which are impacted significantly by climate change. The yield and nutritional quality of pasture are primarily affected by enhancement of atmospheric CO_2 (CO_2 fertilization), rise in temperature, and fluctuations of rainfall pattern. Globally, 200 to 500 million people are practicing the pastoralism either in form of agro-pastoralists, nomadic communities, or transhumant herders, covering around 75% of countries of the world. However, in last four decades, the reports (in journals, conference document, symposium abstracts, popular articles, etc.) showed a decreasing trend of livestock numbers, animal health and productivity, water and food availability, the extent of rangelands, livestock mobility, along with increasing land degradation and overgrazing in pastoral systems throughout the world [18, 50].

As, for example, pastoral productivity has reduced by 20–30% in the second half of the twentieth century in Mongolia [6]. Climate change effects on Mongolia mountain steppes could also be due to a change in land use [71]. During 1961–2013, the lower productivity of livestock and pastoral systems in Africa was due to climate change consequences. Interaction of grassland composition with changed grazing patterns with climate change is an important issue that is needed to be studied elaborately. Above all, few shreds of evidence of an increase in grassland fires in tropical Asia and China are also noticed in the recent past [26] (Fig. 2.4).

2.4.2 Predictions

The enhanced warming would increase the lignification in tissues of fodder, and hence decrease the digestibility of the feed of livestock. Warming-induced enhanced water scarcity decreases the quality and yield of fodder that could also negatively impact the nutrition of livestock. Further, there may be an outbreak of vector-borne diseases of livestock because of changes in precipitation pattern in the cooler region of the world. And also the energy requirement and the demand for water and housing would be accelerated for livestock in future global warming scenarios as higher projected milk demands of the burgeoning population.

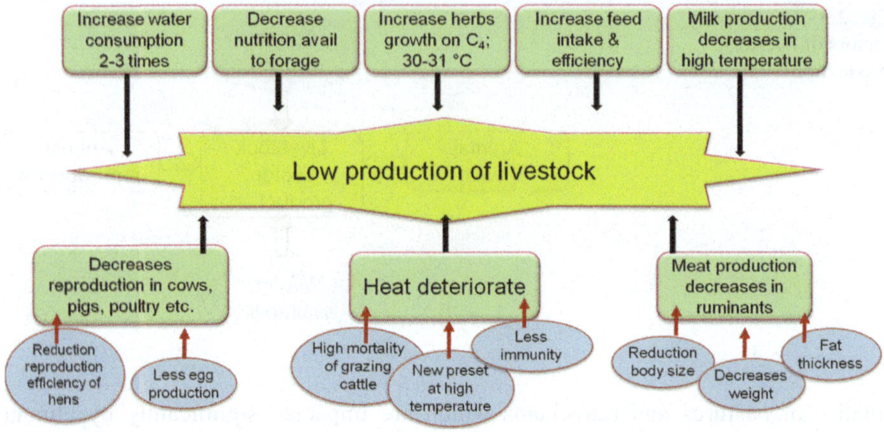

Fig. 2.4 The causes and mechanisms of low production of livestock due to climate change consequences

The pastoral systems would be directly affected by climate change [14]. The majority of other industrial components of livestock systems would suffer indirectly due to rise in the costs of feeding, water, transport, and housing. All of those could lead to uncertainty in the production and volatility of the market prices [49, 63]. The extreme climatic event could also lead to the destruction of the infrastructure of livestock housing and feeding systems.

The projected impacts of climate change (elevated air temperature and CO_2) on pastoral systems primarily focused on herbage growth and quality, and changes in pasture composition [32]. High temperatures with droughts are the critical predisposing factor for devastating grassland fire [26]. The net primary productivity of grazing and grasslands may or may not be affected as there are uncertainties which exit in various projection models [13, 19, 25]. The average annual net primary production (NPP) in rangelands would reduce by 10 g carbon m^{-2} yr^{-1} globally, under the RCP 8.5 scenario in 2050. However, the herbaceous NPP may increase marginally by 3 g C m^{-2} yr^{-1} in a similar scenario [10]. Similar kind of prediction stated that an increase of 2 °C in arid and semi-arid regions could negatively impact the pasture as well as livestock production, while positive impact could be noticed in humid-temperate regions of the world [30, 64]. Not only the productivity but the rangeland composition is also projected to alter [10]. Across rangelands, in southern Tibetan plateau, eastern Great Plains, parts of southern Africa, and eastern Australia, the percentage of the bare ground cover is likely to be increased at a tune of 2.4 times. At the same time, the herbaceous cover would also decline in the eastern Great Plains and Tibetan Plateau. Specifically, the rangelands in southwestern and western parts in Iran could be more vulnerable to future climate change-induced warming-drying situations [42, 67]. Similarly, the shrub covers in rangelands of Tibetan Plateau, eastern Great Plains, eastern Australia, parts of southern Africa, and the Middle East are in threat under climate change scenarios. However, it may increase in some parts of Africa and the Arctic region. The semi-arid southern Sahara might gain both trees and shrub covers in the coming century.

The short-term as well as long-term impacts of climate change on aquaculture are expected. The impact could be direct or indirect. Short-term production losses might be due to extreme heat or cold stresses, high-frequency tropical cyclones, frequent floods, disease outbreaks, the emergence of unexpected parasites, and/or toxic algae [5]. On the other hand, the long-term impacts may be due to scarcity of freshwater for farming as a result of reduced rainfall, less availability and access to marine and terrestrial feeds, scarcity of wild seed, eutrophication, and decreased productivity due to unprecedented extreme climatic events.

Specifically, in India, the performance of dairy animals of northeast and central India is likely to be adversely affected due to heat stress. The sea-level rise in the western coast of India is going to affect the fish breeding, fish migration, and harvest. Moreover, high-frequency extreme cyclonic events in both west and east coast of India in climate change scenarios are going to affect production, capture as well as the marketing of marine fishes negatively. Adding to the damage, the coral bleaching is likely to be increased due to a rise in sea-surface temperature.

The critical point is to mention here that the declines of both fisheries and crop production systems are projected in many countries in the world [7], which would affect the livestock sectors as well as aquaculture simultaneously. The more important issue is to assess the interconnected vulnerabilities of climate change and food systems in developing countries having low HDI (Human Development Index), technical skills, and trade opportunities in the aquaculture sector.

2.5 Impacts of Climate Change on Soil and Water Resources

Soil organic carbon (SOC), the backbone of soil health, is going to be reduced both in the tropical and temperate regions of the world in projected climate change scenarios. Soil health and quality would be negatively affected due to the lowering of SOC status. Further, the plants are going to be exposed in a progressively higher atmospheric CO_2 environment in predicted climate change conditions which would produce plant litters having a higher C: N ratio. The addition of plant biomass having more C: N ratio into the soil will lead to higher carbon immobilization. Consequently, the higher immobilization of SOC reduces the soil nutrient supply to crop. At the same time, the higher nitrogen mineralization due to high soil temperature and quick wet-dry cycle coupled with lower supply would hamper crop productivity which is in the card in future climate change scenarios. Apart from that, the quick mineralization of nitrogen (N) very often leads to higher losses either through volatilization (as NH_3) or by denitrification (as N_2O emission) to the atmosphere which would further cause the positive feedbacks to climate change.

If we take the specific case, then there is a high probability of reduction of SOC in the majority of agro-climatic zones in India. The hilly regions are vulnerable to erosions (due to unprecedented high-frequency rainfall and glacier meting), whereas the coastal regions (Indian-Sundarban, Bengal and Odisha coast along with coastal regions of Andaman and Nicobar Islands) are vulnerable to sea-level rise in predicted climate change scenarios. The lowering of groundwater table in parts of NW and central India due to over-exploitation coupled with higher atmospheric temperature (higher evapotranspiration) is the indirect negative consequences of climate change. However, as predicted, there could be an increase of water flows in *Ganges* and *Brahmaputra* rivers due to the melting of glaciers in the short-term scale but in long run, there would be a scarcity of availability of good quality water.

2.6 Impacts of Climate Change on Food Security

The food as well as nutritional security is under threat in climate change consequences. The four interlinked aspects of food security, namely, food availability, food access, food utilization, and food stability (Fig. 2.5) would be disturbed. How food production could be affected is already described in Sect. 2.2 and 2.3 in this chapter. Further, food access could be limited through disrupting manufacture, transport, and retail that negatively affect the food chain supply. The cycle is vicious (Fig. 2.6). One way the reducing food availability due to climate change consequences is going to increase food costs. Higher food cost leads to higher prices and reduced purchasing power of huge low-income consumers globally [56, 72]. On the other way, the higher food prices decline consumer demand that not only reduce energy intake (calories) but also likely to lead less healthy diet globally [29, 56, 72, 28]. Less healthy diet refers to lower intake of protein and micronutrients (zinc, iron, manganese, etc.) to a huge number of populations in lower and middle-income countries that would increase diet-related mortality [72].

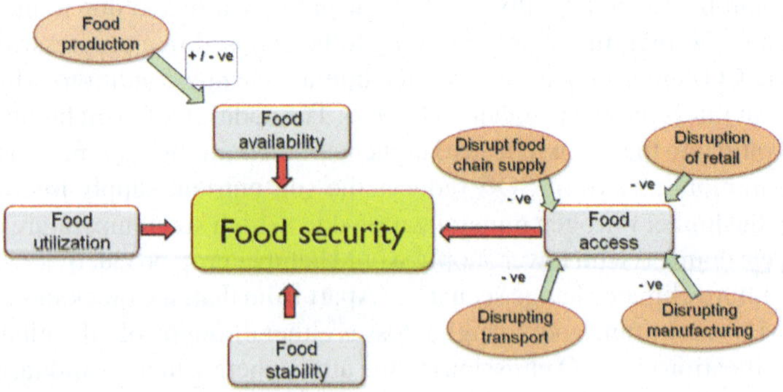

Fig. 2.5 Components of food security

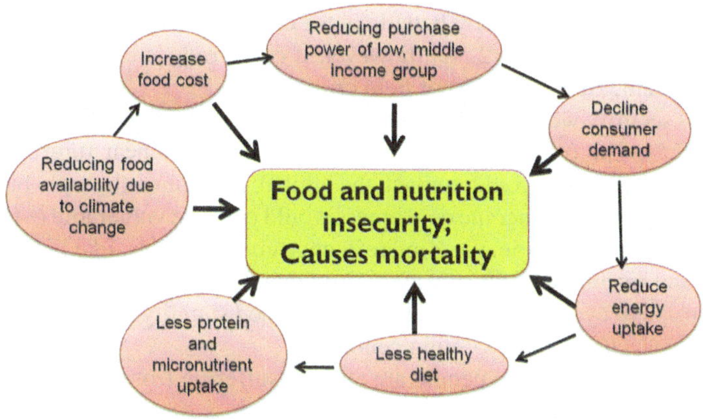

Fig. 2.6 Vicious cycle of climate change consequences and food insecurity

The food quality is directly affected by CO_2 fertilization and increasing/decreasing temperatures that would change the basic metabolism of plants. For example, apples in Japan are blooming earlier due to continuous exposure (for the last three to four decades) to higher temperatures. Consequently, the quality, acidity, firmness, and moisture contents of apples have been deteriorated [73]. Climate change-induced warming could alter the sugar composition, color, and aroma of other fruits like grape, peach, and pear [15]. Not only crops, the unprecedented heat stress is responsible for poor meat (fat and chemical deposition) and egg (shell quality) quality in poultry [45].

The protein content of fruits, vegetables, and field crops is negatively affected by CO_2 fertilization. Meta-analysis showed that around 40% increase in atmospheric CO_2 concentration (~from 390 to 560 ppm) could reduce the zinc and iron content of wheat grain by 9.3% and 5.1%, respectively. The protein content of wheat and rice can also be decreased up to 6.3 and 7.8%, respectively, due to elevated carbon dioxide. However, the changes in nutrients contents were non-significant in the case of C4 crops, field pea, and soybean [55]. An exhaustive meta-analysis on the effect of elevated CO_2 on the nutrient content of different rice cultivars revealed that the average protein, iron, and zinc concentrations were reduced by 10, 8, and 5%, respectively. Moreover, the vitamin contents were also reduced. The vitamin-B1, vitamin-B2, vitamin-B5, and vitamin-B9 were decreased by 17, 17, 13, and 30%, respectively. However, the vitamin E content of rice grain was increased [83]. In contrast, legume and vegetable yields have been found enhanced by 22% under CO_2 fertilization and no significant impact on iron, vitamin C (flavonoid) contents was noticed [69].

The mechanisms of reduced protein content in cereals due to CO_2 fertilization are (i) "carbohydrate dilution" or "growth dilution," i.e., reduction of nitrogen concentration caused by relatively lower nitrogen uptake as compared to biomass growth and (ii) by inhibition of photorespiration that actually could provide much of the energy needed for assimilating nitrate (NO_3^{-1}) into proteins [4, 8, 23]. In a nutshell, the CO_2 fertilization is generally beneficial for crop productivity (mostly C3 plants) at lower temperatures rise (<2 °C); however, it could reduce the nutritional quality (e.g., less protein, zinc, and iron) of field crops.

The food security would be affected by projected future climate change scenarios (RCP 6.0, 8.5). The price of cereals would increase in the near future (2050) due to climate change that would impact the global consumer through higher food prices. Low- and middle-income consumers would be more vulnerable (~183 million people globally). As a whole, the climate change consequences, like increase in both frequency and magnitude of heat waves, droughts, inland and coastal flooding, sea-level rise, tropical cyclone, etc., would decrease food stability and thus increases the risk of food insecurity in the coming century.

Globally, around 2 billion people are already suffering from nutritional deficiencies (protein, micronutrients, etc.) which cause a loss of around 63 million lives annually. And the majority of those people are dependent on C3 grain and legumes for their primary dietary source (iron and zinc) [21, 55]. Specifically, zinc deficiency is presently most prevalent globally and those people who primarily depend on field crops for their dietary needs of zinc are at higher risk [54]. Further, an additional 138 million people are estimated to be vulnerable by 2050 in predicted climate change scenarios. Be specific, the lower and middle-income group of Africa and South Asia is going to be affected most. And projection says, among them, nearly 48 million are residing in India alone [55]. The unfortunate consequence is that by 2050 these negative impacts on protein availability might make a huge population (around 150 million people) vulnerable to protein deficiency [52].

2.7 Messages

The impacts of climate change on agriculture would vary depending on global warming intensity, sea-level rise, CO_2 concentrations, soil-fertility levels, and region. Overall, the total land area including presently highly fertile/productive and waste-lands would be the same in 2050 as today. If we see from a different angle, i.e., cause and consequences of food security and climate change; then the current food system contributes ~40% of total greenhouse gas (GHG) emissions. The food system includes field crops, horticulture, livestock, pastures, rangelands, aquaculture, fisheries, land-use change, deforestation, and wetland degradation; however, the effective food supply system through higher energy/water/fertilizer use efficiencies, reduction of food loss and waste would reduce the GHG emissions and at the same time could improve food security. Last but not least, the package and practices that provide synergies between adaptation and mitigation to climate change would lead to ensuring "climate-smart agriculture" for better health and food security.

References

1. Abiona BG, Fakoya EO, Esun J (2016) The impacts of climate change on the livelihood of arable crop farmers in Southwest, Nigeria. In: Innovation in climate change adaptation. Springer, Cham, pp 289–296
2. Asseng S, Ewert F, Martre P, Rötter RP, Lobell DB, Cammarano D et al (2015) Rising temperatures reduce global wheat production. Nat Clim Change 5(2):143
3. Auffhammer M, Schlenker W (2014) Empirical studies on agricultural impacts and adaptation. Energy Econ 46:555–561
4. Bahrami H, De Kok LJ, Armstrong R, Fitzgerald GJ, Bourgault M et al (2017) The proportion of nitrate in leaf nitrogen, but not changes in root growth, are associated with decreased grain protein in wheat under elevated [CO_2]. J Plant Physiol 216:44–51
5. Barange M, Bahri T, Beveridge MC, Cochrane KL, Funge-Smith S, Poulain F (2018) Impacts of climate change on fisheries and aquaculture: synthesis of currrent knowledge, adaptation and mitigation options. FAO
6. Batima P, Bold B, Sainkhuu T, Bavuu M (2008) Adapting to drought, zud and climate change in Mongolia's rangelands. Clim Change Adapt 197–210
7. Blanchard JL, Watson RA, Fulton EA, Cottrell RS, Nash KL, Bryndum-Buchholz A et al (2017) Linked sustainability challenges and trade-offs among fisheries, aquaculture and agriculture. Nat Ecol Evol 1(9):1240–1249
8. Bloom AJ, Burger M, Kimball BA, Pinter PJ Jr (2014) Nitrate assimilation is inhibited by elevated CO_2 in field-grown wheat. Nat Clim Change 4(6):477–480
9. Bhattacharyya P, Munda S, Dash PK (2019) Climate change and greenhouse gas emission. New India Publishing Agency, New Delhi, 110088. ISBN: 978-93-895-7175-2
10. Boone RB, Conant RT, Sircely J, Thornton PK, Herrero M (2018) Climate change impacts on selected global rangeland ecosystem services. Glob Change Biol 24(3):1382–1393
11. Certini G, Scalenghe R (2011) Anthropogenic soils are the golden spikes for the Anthropocene. Holocene 21(8):1269–1274
12. Chabejong NE (2016) A review on the impact of climate change on food security and malnutrition in the sahel region of cameroon. In: Climate change and health. Springer, Cham, pp 133–148
13. Chen Y, Tao Y, Cheng Y, Ju W, Ye J, Hickler T et al (2018) Great uncertainties in modeling grazing impact on carbon sequestration: a multi-model inter-comparison in temperate Eurasian Steppe. Environ Res Lett 13(7):075005
14. Dasgupta P, Goswami R, Ali M, Chakraborty S, Saha S (2015) Multifunctional role of integrated farming system in developing countries. Int J Bio-Resour Stress Manage 6(3):424–432
15. De Orduna RM (2010) Climate change associated effects on grape and wine quality and production. Food Res Int 43(7):1844–1855
16. Deryng D, Conway D, Ramankutty N, Price J, Warren R (2014) Global crop yield response to extreme heat stress under multiple climate change futures. Environ Res Lett 9(3):034011
17. Din K, Tariq S, Mahmood A, Rasul G (2014) Temperature and precipitation: GLOF triggering indicators in Gilgit-Baltistan, Pakistan. Pakistan J Meteorol 10(20)
18. Egeru A (2016) Climate risk management information, sources and responses in a pastoral region in East Africa. Clim Risk Manage 11:1–14
19. Erb KH, Lauk C, Kastner T, Mayer A, Theurl MC, Haberl H (2016) Exploring the biophysical option space for feeding the world without deforestation. Nat Commun 7:11382
20. FAO (2008) Agriculture organization (2009). The state of food insecurity in the world
21. FAO (2013) FAO statistical yearbook: world food and agriculture. FAO
22. Faye B, Webber H, Diop M, Mbaye ML, Owusu-Sekyere JD, Naab JB, Gaiser T (2018) Potential impact of climate change on peanut yield in Senegal, West Africa. Field Crops Res 219:148–159
23. Feng Z, Rütting T, Pleijel H, Wallin G, Reich PB, Kammann CI et al (2015) Constraints to nitrogen acquisition of terrestrial plants under elevated CO_2. Glob Change Biol 21(8):3152–3168

24. Fenghua S, Xiuqun Y, Shuang L (2006) The contrast analysis on the average and extreme temperature trend in Northeast China. Scientia Meteorologica Sinica 26(2):157–163
25. Fetzel T, Havlik P, Herrero M, Erb KH (2017) Seasonality constraints to livestock grazing intensity. Glob Change Biol 23(4):1636–1647
26. Intergovernmental Panel on Climate Change, Field CB, Barros V, Stocker T, Qin D, Dokken D, Ebi K et al (2012) Managing the risks of extreme events and disasters to advance climate change adaptation. A special report of working groups I and II of the intergovernmental panel on climate change, vol 30(11). Cambridge University Press, Cambridge, UK and New York, NY, USA, pp 7575–7613
27. Gupta R, Wohar M (2017) Forecasting oil and stock returns with a Qual VAR using over 150 years off data. Energy Econ 62:181–186
28. Hasegawa T, Fujimori S, Havlík P, Valin H, Bodirsky BL, Doelman JC et al (2018) Risk of increased food insecurity under stringent global climate change mitigation policy. Nat Clim Change 8(8):699–703
29. Hasegawa T, Fujimori S, Takahashi K, Masui T (2015) Scenarios for the risk of hunger in the twenty-first century using shared socioeconomic pathways. Environ Res Lett 10(1):014010
30. Havlík P, Valin H, Gusti M, Schmid E, Leclère D, Forsell N et al (2015) Climate change impacts and mitigation in the developing world: an integrated assessment of the agriculture and forestry sectors
31. He L, Asseng S, Zhao G, Wu D, Yang X, Zhuang W, Yu Q (2015) Impacts of recent climate warming, cultivar changes, and crop management on winter wheat phenology across the Loess Plateau of China. Agric For Meteorol 200:135–143
32. Herrero M, Henderson B, Havlík P, Thornton PK, Conant RT, Smith P et al (2016) Greenhouse gas mitigation potentials in the livestock sector. Nat Clim Change 6(5):452–461
33. Hooke RL, Martín-Duque JF, Pedraza J (2012) Land transformation by humans: a review. GSA Today 22(12):4–10
34. Huang J, Ji F (2015) Effects of climate change on phenological trends and seed cotton yields in oasis of arid regions. Int J Biometeorol 59(7):877–888
35. Hussain A, Rasul G, Mahapatra B, Tuladhar S (2016) Household food security in the face of climate change in the Hindu-Kush Himalayan region. Food Secur 8(5):921–937
36. Ifeanyi-obi CC, Togun AO, Lamboll R (2016) Influence of climate change on cocoyam production in Aba agricultural zone of Abia State, Nigeria. In: Innovation in climate change adaptation. Springer, Cham, pp 261–273
37. Iizumi T, Ramankutty N (2016) Changes in yield variability of major crops for 1981–2010 explained by climate change. Environ Res Lett 11(3):034003
38. Iizumi T, Furuya J, Shen Z, Kim W, Okada M, Fujimori S et al (2017) Responses of crop yield growth to global temperature and socioeconomic changes. Sci. Rep. 7(1):1–10
39. Innes PJ, Tan DKY, Van Ogtrop F, Amthor JS (2015) Effects of high-temperature episodes on wheat yields in New South Wales, Australia. Agric For Meteorol 208:95–107
40. Intergovernmental Panel on Climate Change (2018) Global warming of 1.5° C: An IPCC special report on the impacts of global warming of 1.5° C above pre-industrial levels and related global greenhouse gas emission pathways, in the context of strengthening the global response to the threat of climate change, sustainable development, and efforts to eradicate poverty. Intergovernmental Panel on Climate Change
41. Intergovernmental Panel on Climate Change (2014) Mitigation of climate change. Contribution of working group III to the fifth assessment report of the intergovernmental panel on climate change
42. Jaberalansar Z, Tarkesh M, Bassiri M, Pourmanafi S (2017) Modelling the impact of climate change on rangeland forage production using a generalized regression neural network: a case study in Isfahan Province Central Iran. J. Arid Land 9(4):489–503
43. Jha B, Tripathi A (2017) How susceptible Is India's food basket to climate change? Social Change 47(1):11–27
44. Ketiem P, Makenzi PM, Maranga EK, Omondi PA (2017) Integration of climate change information into drylands crop production practices for enhanced food security: A case study of Lower Tana Basin in Kenya

45. Lara LJ, Rostagno MH (2013) Impact of heat stress on poultry production. Animals 3(2):356–369
46. Levis S, Badger A, Drewniak B, Nevison C, Ren X (2018) CLMcrop yields and water requirements: avoided impacts by choosing RCP 4.5 over 8.5. Clim Change 146(3–4), 501–515
47. Liu H, Mi Z, Lin LI, Wang Y, Zhang Z, Zhang F et al (2018) Shifting plant species composition in response to climate change stabilizes grassland primary production. Proc Natl Acad Sci 115(16):4051–4056
48. Liu J, Kuang W, Zhang Z, Xu X, Qin Y, Ning J, Wu S (2014) Spatiotemporal characteristics, patterns, and causes of land-use changes in China since the late 1980s. J Geog Sci 24(2):195–210
49. Lopez-i-Gelats F (2014) Impacts of climate change on food availability: livestock. Global environmental change. Springer, Dordrecht, Heidelberg, New York, London, pp 689–694
50. López-i-Gelats F, Fraser ED, Morton JF, Rivera-Ferre MG (2016) What drives the vulnerability of pastoralists to global environmental change? A qualitative meta-analysis. Glob Environ Change 39:258–274
51. Manzoor M, Bibi S, Manzoor M, Jabeen R (2013) Historical analysis of flood information and impacts assessment and associated response in Pakistan (1947–2011). Res J Environ Earth Sci 5(3):139–146
52. Medek DE, Schwartz J, Myers SS (2017) Estimated effects of future atmospheric CO_2 concentrations on protein intake and the risk of protein deficiency by country and region. Environ Health Perspect 125(8):087002
53. Moore FC, Lobell DB (2015) The fingerprint of climate trends on European crop yields. Proc Natl Acad Sci 112(9):2670–2675
54. Myers SS, Wessells KR, Kloog I, Zanobetti A, Schwartz J (2015) Effect of increased concentrations of atmospheric carbon dioxide on the global threat of zinc deficiency: a modelling study. Lancet Global Health 3(10):e639–e645
55. Myers SS, Zanobetti A, Kloog I, Huybers P, Leakey AD, Bloom AJ et al (2014) Increasing CO_2 threatens human nutrition. Nature 510(7503):139–142
56. Nelson R, Kokic P, Crimp S, Martin P, Meinke H, Howden SM et al (2010) The vulnerability of Australian rural communities to climate variability and change: Part II—Integrating impacts with adaptive capacity. Environ Sci Policy 13(1):18–27
57. Onyeneke RU, Igberi CO, Uwadoka CO, Aligbe JO (2018) Status of climate-smart agriculture in southeast Nigeria. GeoJournal 83(2):333–346
58. Rasul G (2010) The role of the Himalayan mountain systems in food security and agricultural sustainability in South Asia. Int J Rural Manage 6(1):95–116
59. Rasul G, Neupane N, Hussain A, Pasakhala B (2019) Beyond hydropower: towards an integrated solution for water, energy and food security in South Asia. Int J Water Resour Dev 1–25
60. Reyer CP, Adams S, Albrecht T, Baarsch F, Boit A, Trujillo NC et al (2017) Climate change impacts in Latin America and the Caribbean and their implications for development. Reg Environ Change 17(6):1601–1621
61. Reyer CP, Bathgate S, Blennow K, Borges JG, Bugmann H, Delzon S et al (2017) Are forest disturbances amplifying or canceling out climate change-induced productivity changes in European forests? Environ Res Lett 12(3):034027
62. Rippke U, Ramirez-Villegas J, Jarvis A, Vermeulen SJ, Parker L, Mer F et al (2016) Timescales of transformational climate change adaptation in sub-Saharan African agriculture. Nat Clim Change 6(6):605–609
63. Rivera Ferre MG, Lópezi Gelats F, Howden M, Smith P, Morton JF, Herrero M (2016) Re-framing the climate change debate in the livestock sector: mitigation and adaptation options. Wiley Interdisc Rev Clim Change 7(6):869–892
64. Rojas-Downing MM, Nejadhashemi AP, Harrigan T, Woznicki SA (2017) Climate change and livestock: Impacts, adaptation, and mitigation. Clim Risk Manage 16:145–163
65. Rosenfeld A, Dorman M, Schwartz J, Novack V, Just AC, Kloog I (2017) Estimating daily minimum, maximum, and mean near surface air temperature using hybrid satellite models across Israel. Environ Res 159:297–312

66. Rosenzweig C, Elliott J, Deryng D, Ruane AC, Müller C, Arneth A et al (2014) Assessing agricultural risks of climate change in the 21st century in a global gridded crop model inter comparison. Proc Natl Acad Sci 111(9):3268–3273
67. Saki M, Esfahani MT, Soltani S (2018) A scenario-based modeling of climate change impacts on the aboveground net primary production in rangelands of central Iran. Environ Earth Sci 77(19):670
68. Saxena AK, Yadav AN, Rajawat MVS, Kaushik R, Kumar R, Kumar M et al (2016) Microbial diversity of extreme regions: an unseen heritage and wealth. Ind J Plant Genet Resour 29(3):246–248
69. Scheelbeek PF, Bird FA, Tuomisto HL, Green R, Harris FB, Joy EJ et al (2018) Effect of environmental changes on vegetable and legume yields and nutritional quality. Proc Natl Acad Sci 115(26):6804–6809
70. Shrestha RP, Nepal N (2016) An assessment by subsistence farmers of the risks to food security attributable to climate change in Makwanpur Nepal. Food Secur 8(2):415–425
71. Spence LA, Liancourt P, Boldgiv B, Petraitis PS, Casper BB (2014) Climate change and grazing interact to alter flowering patterns in the Mongolian steppe. Oecologia 175(1):251–260
72. Springmann M, Godfray HCJ, Rayner M, Scarborough P (2016) Analysis and valuation of the health and climate change cobenefits of dietary change. Proc Natl Acad Sci 113(15):4146–4151
73. Sugiura T, Ogawa H, Fukuda N, Moriguchi T (2013) Changes in the taste and textural attributes of apples in response to climate change. Sci Rep 3:2418
74. Tao F, Zhang Z, Xiao D, Zhang S, Rötter RP, Shi W et al (2014) Responses of wheat growth and yield to climate change in different climate zones of China, 1981–2009. Agric For Meteorol 189:91–104
75. Tiwari PC, Joshi B (2012) Environmental changes and sustainable development of water resources in the Himalayan headwaters of India. Water Resour Manage 26(4):883–907
76. Tripathi A, Tripathi DK, Chauhan DK, Kumar N, Singh GS (2016) Paradigms of climate change impacts on some major food sources of the world: a review on current knowledge and future prospects. Agr Ecosyst Environ 216:356–373
77. Waha K, Krummenauer L, Adams S, Aich V, Baarsch F, Coumou D et al (2017) Climate change impacts in the Middle East and Northern Africa (MENA) region and their implications for vulnerable population groups. Reg Environ Change 17(6):1623–1638
78. Wang P, Zhang Z, Song X, Chen Y, Wei X, Shi P, Tao F (2014) Temperature variations and rice yields in China: historical contributions and future trends. Clim Change 124(4):777–789
79. Ward PS, Florax RJ, Flores-Lagunes A (2014) Climate change and agricultural productivity in Sub-Saharan Africa: a spatial sample selection model. Eur Rev Agric Econ 41(2):199–226
80. Zhang G, Xiao X, Biradar CM, Dong J, Qin Y, Menarguez MA, Doughty RB (2017) Spatiotemporal patterns of paddy rice croplands in China and India from 2000 to 2015. Sci Total Environ 579:82–92
81. Zhao C, Liu B, Piao S, Wang X, Lobell DB, Huang Y et al (2017) Temperature increase reduces global yields of major crops in four independent estimates. Proc Natl Acad Sci 114(35):9326–9331
82. Zhao C, Piao S, Wang X, Huang Y, Ciais P, Elliott J et al (2016) Plausible rice yield losses under future climate warming. Nat Plants 3(1):1–5
83. Zhu C, Kobayashi K, Loladze I, Zhu J, Jiang Q, Xu X et al (2018) Carbon dioxide (CO_2) levels this century will alter the protein, micronutrients, and vitamin content of rice grains with potential health consequences for the poorest rice-dependent countries. Sci Adv 4(5):eaaq1012

Chapter 3
Dimensions of Climate-Smart Agriculture

3.1 Introduction

Climate-smart agriculture (CSA) is an integrated approach considering technical, policy, and investment strategies together in order to achieve food security under climate change conditions [5]. It is relatively a new approach to develop the technical, political, and financial conditions for the achievement of sustainable development goals [4]. There are three pillars of CSA, namely, sustainable productivity, building resilience, and reducing carbon emissions and having six dimensions to build the "three pillars." So, precisely, FAO has forged the concept of climate-smart agriculture (CSA) to improve food security under climate change scenarios.

3.2 Weather-Smart CSA

The components of the "weather-smart" dimension are real-time monitoring and reporting of weather parameters, weather-based crop agro-advisory (CA) services, real-time weather-fluctuation-based crop insurance (CI), climate awareness, and climate-smart housing for livestock.

Continuous monitoring of basic weather parameters related to agriculture like temperature, rainfall, and relative humidity is essential for any sort of climate-smart agriculture. In this connection, establishment of "rain gauge" in each village is a prerequisite. The installation and maintenance of automatic weather station (AWS) at block level could make the system more reliable for giving agro-advisory services and planning of weather-based crop insurance schemes which are the important components of CSA (Table 3.1). The climate information-based real-time value-added agro-advisories to the farmers by state/center/private agencies is important. Policies of CSA must look into crop-specific insurance to farmers in order to compensate for the income losses by them due to weather fluctuation/climate change vagaries.

© The Editor(s) (if applicable) and The Author(s), under exclusive license to Springer Nature Singapore Pte Ltd. 2020

P. Bhattacharyya et al., *Climate Smart Agriculture*, Green Energy and Technology,
https://doi.org/10.1007/978-981-15-9132-7_3

The small and marginal farmers are more vulnerable to climate change consequences. Crops in their fragmented land and livestock of them are very often threatened due to heat and cold stress. Therefore, weather prediction-based protection of livestock from extreme climatic events is an important component of "weather-smart" CSA which is often neglected [3].

3.3 Water-Smart CSA

The major components of "water-smart" CSA can be listed as rainwater harvesting, aquifer recharge, *in situ* moisture conservation, efficient water (irrigation) application system, drainage management, and integrated farming system (Table 3.1).

Rainwater harvesting in watershed either through the construction of water harvesting structures/pond or renovation of the pond is of primary importance. It should be done in series so that the "village water remains in the same village and the farm water remains on the same farm." The runoff networking should be the part of water harvesting model. Effective use of collected runoff in WHS in *rainfed*/dryland areas and its subsequent use in agricultural crops as supplemental irrigation should be encouraged in CSA. Aquifer recharge though "recharging well" or series of "dug-out well" in hilly or plateau would greatly improve the water conservation in dryland areas.

In situ moisture conservation is an age-old concept but-frequently forgotten in water management in agriculture. The laser land leveling, mulching with crop residues, use of hydrogel (site-specific situation only), maintenance of cover crops throughout the year, use of sufficient amount of organic matter/manure in soil, and selective soil amendments are effective technologies which could be followed for *in situ* moisture conservation in agriculture based on their site-specific suitability. On the other hand, reduction of evaporation loss from soil by any means helps in soil moisture conservation.

The sprinkler and drip method of micro-irrigation systems and furrow-irrigated bed planting with proper leveling of land are effective for increasing the application efficiency of irrigation water. In the drip method, water is directly applied to the root zone of crops on demand basis and thereby reduces the water losses. Proper leveling also helps in uniform distribution of water in the crop field and increases the water use efficiency by reducing water losses. Drip irrigation for vegetables and fruit crops already gains impetus throughout the world with higher application efficiencies of 50–60%. Water is going to be a scarce resource in the future and more so in climate change scenarios. Sprinkler irrigation has been a highly successful system of water management of maize in the USA and Brazil. Sprinkler and drip irrigation system should also be considered to other cereals (rice, wheat, etc.) in CSA, where rainfall is limiting and soils are more permeable with region-specific modification in the delivery system. Drainage networking is not only important for lowland ecologies but it is also equally valuable for terrace agriculture in hilly areas. In lowland ecologies, crop productivity, fertilizer application scheduling, reduction of GHG emissions, and weed management are directly dependent on an efficient drainage system. Even the top dressings of fertilizers are not possible in lowland rice if proper drainage is not provided. The methane emission could be reduced in the tune of 35% if one mid-season drainage can be provided in lowland rice-paddy. In hilly upland, diversion drains are essential to maintain the stability of terraces where crops are grown. It -helps in protecting soil erosion in the agricultural-hilly watershed. The integrated farming system (IFS) approach is a unique crop management practice that utilizes the water/moisture efficiently starting from the water harvesting, storing, utilization in farm enterprises, and recycling. The details of IFS, its components, benefits, and potential in CSA are discussed in "Chap. 7" of this book under the heading "Crop management for CSA."

3.4 Crop-Smart CSA

Climate change is a serious threat to food security, agriculture, and livelihood to a huge number of vulnerable population in various places in the world [6]. Extreme weather events, sea-level rise, global warming (increase in temperature), change in rainfall pattern, frequent drought, and floods have negatively affected the crop production [1, 2, 7, 12, 13]. Moreover, the predicted impacts of future climate change on cereal crops showed a significant yield reduction up to −35%, −20%, −50%, −13%, and −60% in case of rice, wheat, sorghum, barley, and maize, respectively [10]. However, yield losses of different regions would vary with the year of prediction and climatic scenarios of the specific regions. The unprecedented decrease in crop biodiversity, low input use efficiencies, and sudden pest outbreaks due to climate change consequences would be the other probable causes of yield losses [8, 14]. Therefore, "crop-smart" adaptation strategies are required to face those challenges to ensure food, livelihood, and environmental security.

Introduction of abiotic stress cultivars, improvement, and adaptation of varieties having higher input use efficiencies, mixed cropping, intercropping, legume-based crop diversification, and changing in planting methods are viable "crop-smart" options of CSA (Table 3.1). Some of them are already practicing but need to be improvised again in site-specific climate change scenarios. Submergence and flood-tolerant rice, terminal heat tolerance wheat, drought-tolerant rice, wheat, maize, etc., cold-tolerant maize, lodging resistant rice-paddy (for cyclonic hazards), frost-tolerant fruits, and vegetable cultivars are to be widely adopted in climate-vulnerable zones. Looking into the nature of future climatic stresses, the multiple stress-tolerant cultivars would be smart approaches. As, for example, drought and submergence tolerance rice cultivars CR Dhan 800 and CR Dhan 801 [9] are fitted for those regions where drought and submergence occur in same growing seasons (either in vegetative stage drought followed by reproductive stages floods or vice versa). The details of stress-tolerant cultivars in major crops are described in Chap. 7 in this book. We are mostly discussing the abiotic stress-tolerant crop cultivars; however, heat-tolerant livestock/fishes should also be a part of CSA. The cultivars having higher nutrient (particularly in case of nitrogen) utilization capacity from native soils or applied fertilizers have to be promoted. The short duration varieties for rice, wheat, maize, and pulses are of great importance in climate hazard perspective. The short duration *rabi* (dry season) varieties (with higher yield potential) not only escape the extreme heat during April, May but at the same time provides alternate windows for planting/sowing in *rainfed* areas during *kharif* season (wet season). The short duration varieties also could be used as contingency crop even after the failure of primary crops due to some extreme climatic events in the same growing season. Mixed cropping and intercropping are old agronomic practice but still have great relevance to CSA. The intelligent cropping systems by incorporating legume or combination of water/nutrient exhaustive and restoring crops together or in the sequence are smart agriculture. The crop diversification is the key approach of "crop-smart" CSA, and it has been also an essential component of conservation agriculture and IFS. The freshwater is a scarce resource and is going to be more limiting in climate change scenarios. The change in planting methods for major cereal crops not only helps in saving water but also reduces the GHG emissions. For example, drydirect seeding (DDSR) as compared to transplanted rice, raised-bed planting of maize/sugarcane, pit plantation for fruits, and coir-bed plantation of vegetables are few "crop-smart" technologies of CSA.

3.5 Nutrient-Smart CSA

The smart management of nutrients is the key strategy of CSA. Ensuring higher production, system resilience, and GHG mitigation, all three are addressed adequately through proper nutrient management. For example, the real-time nitrogen management ensures higher yield of the crop (by supplying demand-based requirement), increases NUE, reduces losses of nitrate (reduces groundwater pollution), and mitigates N_2O emissions simultaneously. The major "nutrient-smart" CSA includes green manuring (GM), brown manuring (BM), site-specific integrated nutrient management (SINM), leaf color chart (LCC)-based nitrogen application, soil test-based nutrient management (SSNM), and soil health card-based nutrient management. Real-time site-specific INM, in general, refers to the optimum supply of nutrients matching with the requirements of crops over space and time with the right product, place, rate, and time. Leaf color chart (LCC) or customized LCC (CLCC)-based nitrogen application has been shown effective in N-management by improving NUE by 5–15% and reducing the N_2O emissions by 7–12% in rice. The LCC base fertilizer management is also found suitable in wheat and maize. The SPAD reading-based N-management in rice and other crops is promising "nutrient-smart" technology that must be promoted in CSA. The sensor-based major- as well as micronutrients delivery systems in precision farming are other options. Green manuring is an age-old practice in rice where *Sesbania* sps. are grown in the field for about 40–45 days and incorporated in the soil before puddling/transplanting of rice which could provide 25–30 kg ha^{-1} nitrogen to soil and maintain soil physical condition. In brown manuring (BM), the standing green manuring crops are knocking-down with the chemical spray (2, 4-D) in absence of standing water in the field. The rotten/dead *Sesbania* crop is then slowly mixed or decomposed in soil and subsequently provides N to the system. The GM is followed where sufficient water is available during transplanting of rice, otherwise BM is a good alternative option. In DSR also GM/BM could be introduced by sowing both the rice and *Dhaincha* (*Sesbania*) seeds together with seed drill and subsequently either incorporated (with "cono-weeder") or "knock-down" by 2, 4-D after 25–30 days of sowing. These practices could provide 15–20 kg nitrogen per hectare and at the same time suppress weed infestation during the initial 20–25 days in DSR which is crucial. "Soil health card"-based nutrient management is an improvement of previously followed "soil test-based nutrient management" in agriculture. It is a climate-smart approach, however, it needs robust institutional support for testing (for soil testing and authentic reporting) and follow-up recommendations. The incorporation of legumes in a cereal-cropping system improves N-supply and soil quality. It is "crop-smart" as well as "nutrient-smart" CSA. In this connection, it is to be remembered that all the dimensions of CSA or smart technologies of CSA are interrelated and interdependent; however, for the shake of understanding, we have categorized them and discussed separately here.

3.6 Carbon and Energy-Smart CSA

The carbon- and energy-smart technologies primarily aim to reduce GHG emissions and thereby mitigating climate change. The "Resource Conservation Technologies (RCTs)," "Conservation Agriculture (CA)," agroforestry, and integrated pest management (IPM) are some "carbon/energy-smart" practices in agriculture (Table 3.1). These practices help in carbon sequestration and reduce CO_2, CH_4, and N_2O emissions in the tune of 15–25% based on the location and crops/livestock used. Residue management and zero tillage which are an integral part of CA are "carbon- and energy-smart" agriculture which has multidirectional benefits including soil carbon buildup, soil compaction, better infiltration, maintenance of optimum soil porosity, microbial health protections, improving soil aggregation, etc. In this connection, use of second-generation farm machineries like "happy-seeder," "zero-tilled seed drill," "straw chopper cum bailer," "combined harvester," "power-operated rice transplanter," "solar-operated weeder," "residue composter," etc. are not only important but necessary also to make the system carbon/energy smart. Fodder management, concentrate feeding for livestock and feeds that inhibit methanogenesis in the rumen are few smart technologies that should be used in livestock management. The details of livestock, aquaculture, and fisheries management in CSA are described in "Chap. 8" of this book. Location-specific agroforestry and sustainable wetland management have huge potential for carbon sequestration that needs to be promoted in CSA. Hilly catchment treatments (with bunds and afforestation), water harvesting in check dams, and using gravitation energy (slope geometry) for irrigation to command areas in hills are energy-smart technologies that need to be encouraged in people participation mode (details are discussed in "Chap. 7" of this book). Reduction in the use of chemicals through IPM modules or nano-based delivery system (low quantity with higher efficiencies) is an "energy-smart" option which could be considered in CSA.

3.7 Institute and Knowledge-Smart CSA

Institutionalization of technology delivery and maintenance systems along with solid knowledge base (an amalgamation of traditional and modern scientific knowledge) are essential to make the agriculture "knowledge smart." Very often, we are talking about the institutional/organizational support to innovative technologies for better adaptation; however, CSA needs to institutionalize the technology innovation cum delivery mechanism. Community-based seed and fodder bank, participatory community nursery for rice, village-climate-risk management committee (VCRMC), CHC, and transparent custom hiring center for farm machineries are few examples of those. (Table 3.1)

The typical "knowledge-smart" CSA refers to gender equality, ICTs, and capacity building of farm youth/farmers through an awareness campaign and onsite training. Mobile-based apps for nutrient, pest, and disease management with a smart advisory in the local language nowadays gained popularity. It is attractive to farm youth and

Table 3.1 Components and technology framework of CSA

S. No.	Weather smart	Water smart	Crop smart	Nutrient smart	Carbon smart	Institute smart
1	Rain gauge	Aquifer recharge	Stress-tolerant varieties	Soil health card	Residue recycling	VCRMC
2	Automated Weather Station (AWS)	In situ moisture conservation	Input-efficient varieties	SSNM	CA	CHC
3	Weather-based agro-advisory	Efficient application system	Inter-cropping	Legumes	Tank silt application	Seed and fodder banks
4	Insurance	Drainage	Diversification	INM	Agroforestry	Community nursery
5	Awareness	IFS	Planting methods	Coated fertilizer	Livestock management	Capacity building

Source [11]

also makes the system transparent that would provide quick real-time solutions with minimum cost. Gender equality and social issues are of equal importance in CSA like other approaches for its better adaptation. The women farmers are the backbone of agriculture (accounting for 43–45% of the farm workforce) and rural economy, and hence the important driver of CSA. Therefore, the CSA also should be "gender smart." The CSA policy must include skill development of women, equal assets sharing, access to credit, increase their communicating ability, and engaging them to decision-making processes related to food security, resilience, and climate change issues. Self-help group (SHG) formation of women, leadership development, and adequate training are the three key factors of "knowledge-smart" CSA .

References

1. Bhattacharyya P, Munda S, Dash PK (2020) Climate change and greenhouse gas emission. New India Publishing Agency, New Delhi, 110088; pp 170+25. ISBN: 978-93-895-7175-2
2. Brida AB, Owiyo T, Sokona Y (2013) Loss and damage from the double blow of flood and drought in Mozambique. Int J Global Warming 5(4):514–531
3. Chhetri AK, Aggarwal PK, Joshi PK, Vyas S (2017) Farmers' prioritization of climate-smart agriculture (CSA) technologies. Agric Syst 151:184–191
4. Food and Agriculture Organization (FAO) of the United Nations (2015) AQUASTATDFO. AQUASTAT, (Database). https://data.fao.org/ref/75f7d9c5-57ab-4a62-88d3-1e47fb50c45
5. Food and Agriculture Organization (FAO) of the United Nations (2013) Climate-smart agriculture-sourcebook. E-ISBN978-92-5-107721-4
6. IPCC (2014) Summary for policymakers. Climate change 2014: impacts, adaptation, and vulnerability. Part A: global and sectoral aspects. Contribution of working group IIto the fifth assessment report of the intergovernmental panel on climate change. Cambridge University Press, Cambridge, UK and New York, NY, USA, pp 1–32
7. Lobell DB, Sibley A, Ortiz-Monasterio JI (2012) Extreme heat effects on wheat senescence in India. Nat Clim Chang 2(3):186–189
8. Norton R (2014) Combating climate change through improved agronomic practices and input-use efficiency. J Crop Improv 28(5):575–618
9. Pathak H, Parameswaran C, Subudhi HN, Prabhukarthikeyan SR, Pradhan SK, Anandan A, Yadav MK, Aravindan S, Pandi GP, Gowda BG, Raghu S (2019) Rice Varieties of NRRI: Yield, Quality, Special Traits and Tolerance to Biotic & Abiotic Stresses
10. Porter JR, Xie L, Challinor AJ, Cochrane K, Howden SM, Iqbal MM, Lobell DB, Travasso MI (2014) Food security and food production systems. pp 485–533
11. Prasad YG, SrinivasaRao CH, Prasad JVNS, Rao KV, Ramana DBV, Gopinath KA, Srinivas I, Reddy BS, Adake R, Rao VUM, Maheswari M, Singh AK, Sikka AK (2015) Technology demonstrations: enhancing resilience and adaptive capacity of farmers to climate variability. National Innovations in Climate Resilient Agriculture (NICRA) Project, ICAR-Central Research Institute for Dryland Agriculture, Hyderabad, p 109
12. Prasanna V (2014) Impact of monsoon rainfall on the total foodgrain yield over India. J Earth Syst Sci 123(5):1129–1145
13. Singh G, Mishra D, Singh K, Parmar R (2013) Effects of rainwater harvesting on plant growth, soil water dynamics and herbaceous biomass during rehabilitation of degraded hills in Rajasthan, India. For Ecol Manage 310:612–622
14. Zabel F, Putzenlechner B, Mauser W (2014) Global agricultural land resources–a high resolution suitability evaluation and its perspectives until 2100 under climate change conditions. PloS ONE 9(9):e107522

Chapter 4
Soil Management for Climate-Smart Agriculture

4.1 Soil Quality, Soil Resilience, and Soil Health

We should understand the concept of soil quality and its assessment, primarily for two reasons, firstly, to check the damage and improper usage of soil system and to match the management options to soil's capability. The other reason is to know the baseline of the capabilities of soil function in order to recognize further changes on a real-time basis. The soil testing is the point data of a particular time which only provides a snapshot of soil fertility status, not its functional capabilities. Routine soil testing not able to identify the deterioration of soil properties leads to production constraints. In this context, soil quality provides a holistic judgment of soil's functional capabilities and management-induced changes.

Concept and definition of soil quality

Soil quality basically enumerates the synergy of soil biological, chemical, and physical processes that enable it to perform critical ecological functions [2, 13]. The functions primarily include maintaining dynamics of nutrients in the soil-plant system, conserving biological diversity and restoring its functions, regulating water-solute flow dynamics, detoxifying and buffering toxic substances, and sustaining socio-economic support system. Soil quality is not merely a measure of certain soil properties but it is a logical assessment of the performance of soil with respect to both in the present situation and also for a future scenario.

Initially, Pierce and Larson [28] defined the soil quality as "fitness for use," later in 1994, Doran and Parkin referred soil quality as the "capacity of soil to function." They gave the comprehensive definition of soil quality as "the capacity of soil to function within boundaries to sustain biological productivity, maintain environmental quality and promote plant and animal health." In 1995, Lal and Stewart [17, 18] described soil quality in terms of its inherent capacity to provide ecosystem services and protecting environmental sustainability. However, 2 years later, Karlen et al. [13, 14] gave a

© The Editor(s) (if applicable) and The Author(s), under exclusive license to Springer Nature Singapore Pte Ltd. 2020

P. Bhattacharyya et al., *Climate Smart Agriculture*, Green Energy and Technology,
https://doi.org/10.1007/978-981-15-9132-7_4

most comprehensive definition of soil quality which stated, it is "the capacity of a specific kind of soil to function within natural or managed ecosystem boundaries, to sustain plant and animal productivity, maintain or enhance water quality and support human health and habitation."

Soil quality assessment

We know that soil quality is an assessment of soil's ecological functions on a time scale. So, it needs to be assessed periodically. The basic purpose of the assessment is to maintain the soil for habitat well-being and to develop water and agricultural productivity. Recent assessment approaches integrate soil characteristics/processes with the performance of functions and management practices. Therefore, assessment of soil quality is three-tier approaches, first to formulate appropriate goals to perform specific soil functions, then selection of minimum dataset (MDS) for selecting the indicators, and lastly developing the soil quality indices through statistical scoring and ranking procedure.

Soil quality indices must be selected on the management goals of the system based on specific site-specific functions (Fig. 4.1). Management goals include on-farm productivity, environmental effects, and societal consequences. It could also consider soil erosion, contamination, and socio-economic issue like subsidy imbalance. However, management goals differ depending on vision of policy makers, land managers and interest of stakeholders. So, the soil functions need to be considered for assessing soil quality indices that must be based on management goals. Therefore, soil functions may be many for fulfilling a single management goal (Fig. 4.1). The second important job for assessing soil quality indices is the selection of indicators. Minimum datasets (MDS) are required for the selection of indicators. The MDS should consider physical, chemical, and biological properties/processes of soil which are easy to measure and could detect changes in soil function, accessible

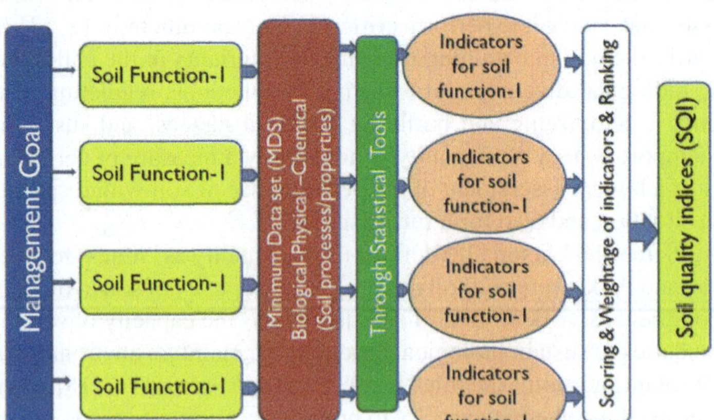

Fig. 4.1 Logical framework of assessment of soil quality and developing soil quality indices (SQI)

to stakeholders, and applicable to field conditions. In the recent time, at the era of process-based simulation modeling (which is used to upscaling and predicting the soil quality indices), MDS should also be sensitive to management practices and climate change [2, 7]. The indicators from MDS could be selected through expert opinion, principle component analysis (PCA), factor analysis, multiple correlations, star plot technique, cluster analysis, etc. In agricultural studies, mostly PCA is used. The majority of the techniques use statistical tools for selecting a large number of representative observations rather than individual preferences. The main function of these selections is to reduce the number of indicators. The indicators thus selected must represent the specific soil function and may be different for different functions. After the indicators are selected, the outputs are transferred to a linear or non-linear scoring technique. In order to develop the soil quality indices (SQI), the indicators are first converted to unit less numerical scores for setting the limits or thresholds for each indicator. The assignment of proper critical limits to indicators for a specific soil function is the key to the success of SQI [21] (Table 4.1). After finalizing the critical limits, the numerical scores of each are transferred to scoring functions. Then in the final step, indicator's scores are integrated into a comparative SQI (Eq. 4.1)

$$SQI = \sum_{i=1}^{n} Wi * Si \tag{4.1}$$

where $i = 1-n$; soil quality indicators; S = Score for the indicators; W = Weighting factors of each indicator, for specific functions representing a definite management goal.

In (Eq. 4.1), the higher SQI values represent the better quality of the soil. The optimum value of SQI could be assigned with the reference to an undisturbed ecosystem considering maximum potential or to the best management system (Fig. 4.1).

Soil resilience

Soil resilience is the ability of soil to resist any change and bounce back (recover) to its original state. Primarily, it refers to the elasticity (the rate of recovery), the amplitude (extent in which recovery possible), and hysteresis (the pattern or path of recovery) of soil (Fig. 4.2). Therefore, we can say soil quality and soil resilience are interrelated but different from each other. Broadly, soil quality addresses the soil functions and ecosystem services, while resilience tells its (soils) ability to restore itself when subjected to stresses. Empirically, soil resilience could be expressed in terms of soil quality (Eq. 4.2) [19].

Table 4.1 Different soil quality indicators with their scoring curve and limits

Indicator	Scoring curve	Lower threshold	Upper threshold	Lower threshold
Physical indicators				
Clay (%)	More is better	0	40	20
Bulk density (Mg/m^3)	Less is better	1	2	1.5
Hydraulic conductivity (cm/h)	Optimum	0.2	2	0.6
Clay dispersion index	Less is better	0	36	18
Physico-chemical indicators				
pH	Optimum	4.5	9	5.5
Electrical conductivity (dS/m)	Less is better	2	12	6
Chemical indicators				
Organic carbon (g/kg)	More is better	0	12	6
Carbon mineralization (mg/kg)	More is better	0	1200	600
Total nitrogen (mg/kg)	More is better	0	1200	600
Available nitrogen (kg/ha)	More is better	0	400	200
Nitrogen mineralization (mg/kg)	More is better	0	60	30
Bray's phosphorus (kg/ha)	More is better	0	50	25
Available potassium (kg/ha)	More is better	0	400	200
DTPA zinc (mg/kg)	More is better	0	1.5	0.75
DTPA copper (mg/kg)	More is better	0	5	2.5
DTPA iron (mg/kg)	More is better	0	50	25
DTPA manganese (mg/kg)	More is better	0	20	10
Biological indicators				
Microbial biomass carbon (mg/kg)	More is better	0	400	200
Microbial biomass nitrogen (mg/kg)	More is better	0	60	30

(continued)

Table 4.1 (continued)

Indicator	Scoring curve	Lower threshold	Upper threshold	Lower threshold
Urease (µg NH_4^+ /g/h)	More is better	0	200	100
Dehydrogenase (µg TPF/g/h)	More is better	0	100	50
Acid phosphatase (µg PNP/g/h)	More is better	0	600	300
Alkaline phosphatase (µg PNP/g/h)	More is better	0	400	200

Sources Lal [15]; Bhattacharyya [2]

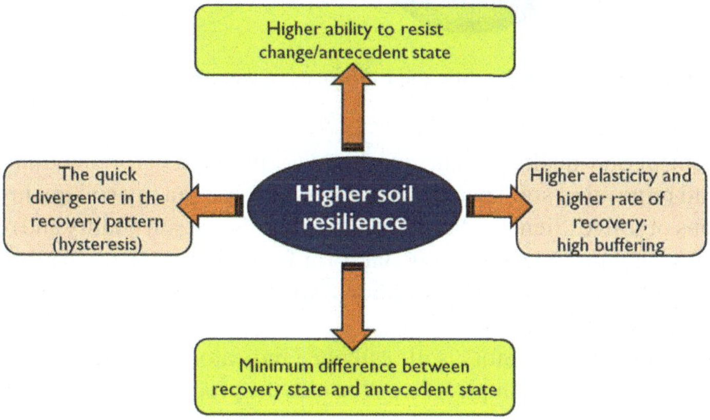

Fig. 4.2 Conceptual framework of soil resilience

$$S_r = S_a + \int_0^t (S_n - S_d + I_m)\, dt \quad (4.2)$$

where S_r = Soil resilience; S_q = Soil quality; t = time; S_a = antecedent soil condition (soil quality); S_n = rate of soil renewal; S_d = rate of soil degradation; and I = management input (positive).

$$S_r = -dS_q/dt \quad (4.3)$$

Soil resilience depends on its inherent properties, climatic conditions prevail in a specific location, water balance (hydrology), biodiversity exist, parent materials prevail, and the terrain of the region. However, soil resilience could be enhanced significantly with efficient land-use management. Efficient and site-specific land management practices enhance soil restoration capacity. Those could be listed as

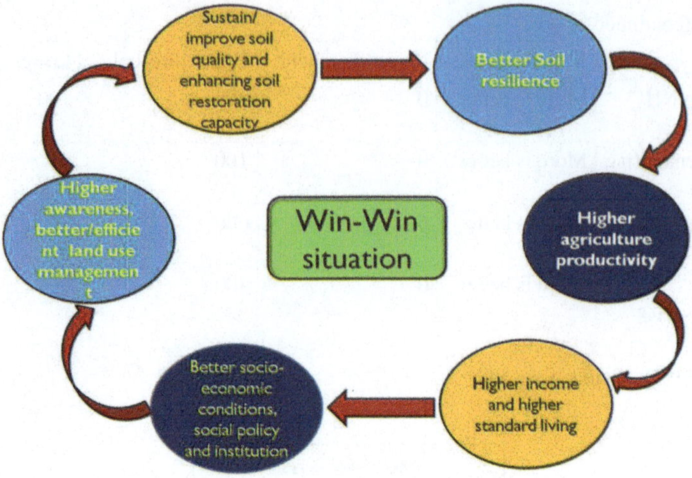

Fig. 4.3 Sustain soil quality and restore soil resilience is a win-win situation

improvement of soil quality, enhancing biomass productivity, reduction of abiotic stresses, and removal of soil-specific constraints. So, we can say soil quality is one of the functions of soil resilience. Therefore, if we can sustain soil quality and restore soil resilience it would provide a win-win platform for higher productivity, better socio-economic stability, and efficient institutional functionality which further encourage higher soil resilience (Fig. 4.3).

Unlike soil quality indicators, soil resilience indicators are different for a certain ecological function. For example, if we consider the productivity and environmental regulation are the functions, then the response to input, buffering capacity, higher SOC accretion rate, higher productive subsoil, uniform deep soil profile, and the response of soil processes to restorative measures could be identified as soil resilience indicators. On the other hand, for the same function, soil depth, erosivity, SOC content, nutrient, and water use efficiency could be the soil quality indicators. So, up to this, we understand that soil resilience is an ecological concept. However, unlike soil quality, the assessment of soil resilience is not easy as it is a relatively long-term phenomenon, and quantifying the recovery rate, elasticity, and hysteresis are tricky in natural dynamics systems, like soil. However, both the qualitative and quantitative methods are followed for assessing soil resilience (Fig. 4.4). Qualitative measures include dynamic soil processes, vegetation diversity/cover, biomass production, and nutrient-energy-water balance. On the other hand, quantitative soil resilience could be determined through process-based simulation models, buffering capacity, and kinetics (rate) of soil restoration.

Soil Health

Soil quality term has widely been accepted throughout the scientific community until criticism started at the end of the 1990s. In 1997, Doube and Gupta [26] coiled the term soil health and defined it as "capture the ecological attributes of the soil which

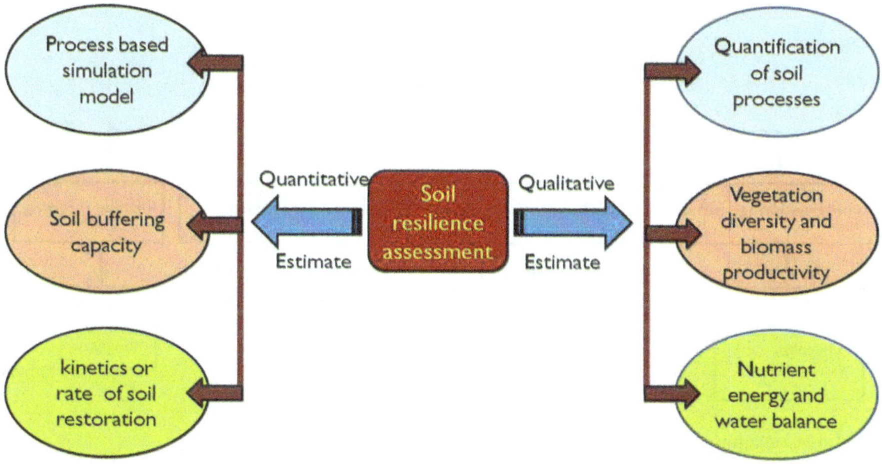

Fig. 4.4 Framework of assessment of soil resilience

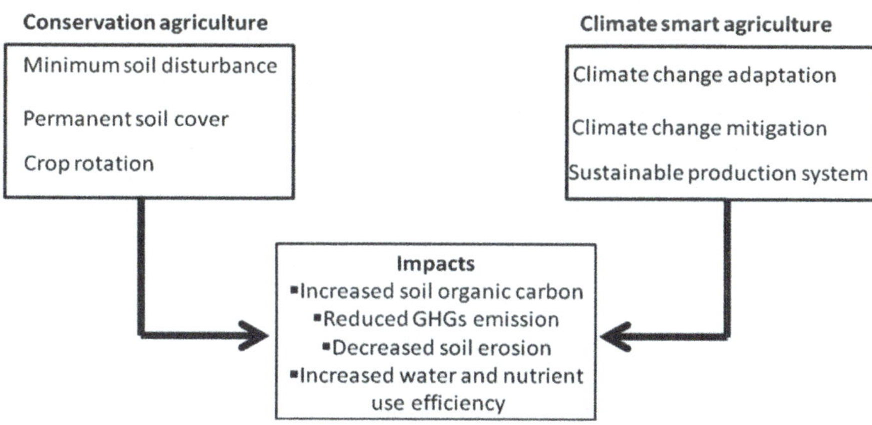

Fig. 4.5 Interlinked between conservation agriculture and CSA approaches

have implications beyond its quality or capacity to produce a particular crop. These attributes are chiefly those associated with the range of functions it performs" [3]. However, the main differences pointed out that the soil health primarily considers the dynamic and living nature of the soil, although very often both the terms are used synonymously. But if we go critically, then soil quality indicates more on soil capacity to produce a particular crop for human need, while soil health focuses on sustaining soil's function. Meanwhile, Moebius-clune, [22] described the soil health is nothing but equivalent to dynamic soil quality. Overall, observations and documentations prefer to use soil quality terms for researchers and soil health term for farmers and policymakers.

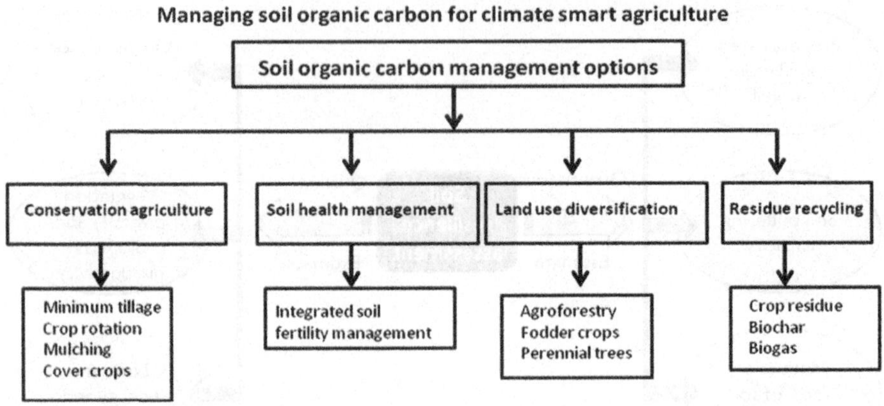

Fig. 4.6 Conceptual framework of managing soil carbon for CSA

4.2 Climate Change and Issues of Soil Sustainability

The influence of climate change and global warming on soil processes/properties is complex and driven by various direct and indirect factors. Enhanced rainfall would increase the surface runoff in hilly terrain, which, in turn, enhances the water storage and infiltration in the valley or plain lands. Increased precipitation also could enhance the groundwater recharge in the permeable and well-drained soils. In contrast, in the relatively warmer region having low soil-infiltrability, higher evapotranspiration from the soil-plant system may cause agricultural drought. The climate-induced rise in temperature would increase the potential evapotranspiration and decrease the water storage, infiltration, surface runoff, and groundwater recharge, particularly in the low rainfall region. The major impacts of climate change on soil and soil sustainability could be listed in seven points as follows:

(i) Reduction of quality and quantity of organic matter, in general, and soil organic carbon, in particular, in tropical and sub-tropical areas.
(ii) Reduction in rate of decomposition of added crop residue with a higher C:N ratio than expected to generate in higher quantity at elevated CO_2 concentration in the atmosphere.
(iii) Addition of crop residues with a higher C:N ratio would reduce nutrient mineralization (increased immobilization) in the soil in anticipated climate change scenarios.
(iv) Nitrogen, phosphorus, and potassium mineralization may increase in higher temperature but their availability would decrease due to higher losses by the processes of denitrification and volatilization.
(v) Soil erosion may increase as the intensity of rainfall would be expected to increase, although total precipitation may be unchanged or decreased.
(vi) Distribution and frequency of rainfall and wind intensity could enhance water and wind erosion leading to soil degradation.

(vii) Sea-level rise may lead to salinity ingression in the coastal productive lands of Asia, America, Africa, and Australia and turning them to less productive agriculture.

4.3 Soil Resources and Principles of Soil Management for CSA

4.3.1 Soil Resources

The climate-smart agriculture (CSA) aims to meet the demand for developing a strategy for sustainable management of food and agriculture systems under the face of global climate change. The CSA specifically aims to achieve three objectives [9, 20], viz., (i) sustainable increase in agricultural productivity to enhance income and food security, (ii) building resilience to climate change from the farm to national levels, and (iii) to control emissions of GHGs from the agriculture sector.

Soil is one of the major terrestrial pools of carbon and globally contains about 2344 Gt of organic carbon [32]. The current estimate says that, on a global scale, soils contribute 37% of the GHG emission from the agricultural sector [33]. In contrast, the agricultural soils could also significantly contribute to GHG mitigation through soil organic carbon (SOC) sequestration [27]. The amount of carbon stored in soil is much more in quantity compared to carbon stock in living vegetation and the atmosphere added together, and therefore any minor change in soil organic carbon can lead to marked alteration in the change in CO_2 concentration in atmosphere [31].

Globally, soils are considered to have a high potential of carbon sequestration, which can be achieved by the adoption of suitable land uses and best soil management options [6]. Emissions of GHGs are strongly affected by land-use change (LUC) and/or soil management. The surface SOC layers are also sensitive to those changes and management practices. Therefore, careful monitoring and management of LUC give an effective opportunity to bring down atmospheric GHG concentrations. Now, to make the soil management practices sustainable, the basic ecosystem functions performed by soil are to be maintained or enhanced without adversely affecting those services or biodiversity [8].

4.3.2 Principles of Soil Management for CSA

Climate change and variability affect soil health in various ways. In general, under changing climatic scenarios, the weather events would be more intense and erratic, and this would lead to increased erosion by water and wind, more runoff loss, reduced groundwater recharge, and the deficit of soil moisture for plant growth. Further, the decreased or erratic rainfall and significant drought periods would cause soil moisture stress, thus limiting water and nutrient availability to plants. Moreover, the increased soil surface temperatures will increase the mineralization of soil organic matter, hinder carbon sequestration capacity, increase salinization, and increase evaporation rates, which will ultimately limit plant growth. Increased salt accumulation in the soil would also restrict plant growth and reduce agricultural productivity.

Therefore, soil management practices should focus on increasing soil organic carbon content. The practices should aim at lowering chemical input to the soil and maintaining productive soil, rich in carbon. The integrated soil management practices for CSA should aim to create optimum soil physical, chemical, and biological conditions with a special focus on soil resilience. Principles and practices of some management practices are as follows:

(i) No or minimum tillage system with retention of crop residue would lead to minimizing soil disturbance and the protection of soil organic carbon.
(ii) Burning of crop residues must be avoided to maintain optimum soil physical and biological conditions.
(iii) In an intensive agriculture system, nutrient supply should be regulated from combined inputs of organic (mulch, compost, crop residues, green manure) and inorganic sources.
(iv) Management practices that inhibit the rapid mineralization of carbon and nitrogen in the soil are to be adopted.
(v) Application of biochar, which is a carbon-enriched form of charcoal, with high stability, porosity, and surface area (produced by pyrolysis, that is, heating of biomass like rice husks, sugar cane bagasse, municipal wastes, and animal manure, etc., with little or no oxygen) could enhance the SOC sequestration potential in soils. Conversion of rice straw into biochar is a potential strategy for controlling GHG emissions [30].

4.3.3 Sustainable Soil and Land Management for CSA

Once the soil is degraded, it becomes more vulnerable to the adverse impact of climate change. Serious losses of soil organic matter take place from degraded soil which further accelerates soil erosion. The sequestration of organic carbon in soil is a key step to CSA. It has significant potential to develop sustainable agricultural production systems and increasing the resilience of ecosystems [8]. It would also help to favor ecosystem services provided by soil. The large rehabilitation of degraded soil is the key to increase soil organic carbon with a simultaneous reduction in soil erosion.

Prevention of land degradation

Large-scale restoration of degraded soils is fundamental for sequestering soil carbon (C) and sustaining the productive capacity of the soil. Increasing soil organic matter and reducing soil erosion are the two basic criteria for promoting soil organic C buildup. The world's soils are seen as high potential for carbon sequestration through appropriate land uses and agricultural management practice. Moreover, degraded soils are much more vulnerable to the negative impacts of faulty management practices and climate change. Land degradation reduces soil organic carbon, biodiversity, and accelerates soil erosion, which all together increases vulnerability to climate change.

On the other hand, unsustainable land management practices promote soil degradation. Some of those practices are tillage-based crop production systems, intensive mono-cropping (resulting in soil nutrient mining), imbalanced fertilizer application, faulty irrigation practices, burning of crop residue, and degradation of forest lands. For example, a 30% decrease in soil-C stock in a period of fewer than 5 years due to unsustainable land-use change has been reported by Zach et al. [34] and Noellemeyer et al. [23]. In an Amazon rainforest, establishing pastures on cleared forest could emit around 8 to 12 tons of CO_2 per hectare land [4, 10]. Therefore, unsustainable soil management practices must be prevented especially in the areas where the soil organic C stocks are inherently less.

Controlling soil erosion

Erosion removes the fertile topsoil which is the most productive part of the soil profile causing losses of soil organic matter. Intensive tillage induces soil erosion, which further accelerates the emission of carbon from the soil. As an alternative to intensive tillage, conservation agriculture (zero-tillage based) has the potential to reduce soil erosion. Minimum- or no-tillage-based farming system in combination with suitable soil and water conservation measures is the effective adaptive measure for reducing erosion. The controlled grazing to sustain the quality of rangeland vegetation, contour farming using bunds and diversion ditches, and planting windbreaks perpendicular to the prevailing winds are the other adaptive options to reduce soil erosion. On steeper slopes, bench terracing, planting across the slope are used to optimize water capture and infiltration. In different parts of the world, biological geotextiles have been used for controlling soil erosion and supporting above-ground biomass production (Fig. 4.5).

4.3.4 Managing Soil Organic Carbon Stock for CSA

Soil fertility and water holding capacity both are improved with an increase in soil carbon sequestration. So, soil carbon sequestration could play an important role in improving plant productivity. The SOC sequestration rate is controlled by several factors such as climate, soils, cropping systems, and agronomic management practices. Specifically, the soil carbon sequestration varies with changing tillage practices in fine-textured soils, while straw incorporation had a greater impact on soil organic carbon compared to tillage in coarse-textured soils. Insufficient addition of organic input in soil may also promote SOC loss. For instance, in the USA, farmland conversion into grasslands was found to be a major reason for moderate increases in SOC [24, 25].

As we know, the heterotrophic respiration contributes to half of the carbon emission, while soil biological activities govern the entire nitrogen (N) emission from soil systems. Soil management practice, especially tillage and cover crops can regulate the emission of biogenic greenhouse gases (GHGs). About one-third of anthropogenic CO_2 emissions since 1750 took place due to land-use changes involving tillage [12]. If historical depletion can be reversed, these soils may sequester up to 50 ppm of atmospheric CO_2 over 50 years, which can decrease 40% of current anthropogenic carbon emission in the atmosphere. It is assumed that the decreasing tillage intensity can increase SOC levels, although another school of thought has proposed that tillage redistributes SOC rather than decreasing SOC [1]. The effect of tillage on soil N_2O emissions is debatable. It has been found that tillage decreases N_2O emissions by reducing soil anoxic conditions. In contrast, after the tillage-induced initial loss of soil carbon, the N_2O emission from N fertilizer significantly contributes to net GWP in many cropping systems. Overall, the soil-C stocks depend on the agro-climatic zone and type and intensity of its use. Soil organic C improves soil structure by binding soil particles into water-stable aggregates, thus improving water holding capacity, infiltration, aeration, and soil nutrient availability. Therefore, soil organic C also has an impact on the biological resilience of agricultural ecosystems. Specifically, increasing SOC content increases soil biodiversity. Higher levels of biodiversity promote the processes involved in breaking down organic matter and releasing nutrients to plant-available forms. Organic and inorganic carbon pools in soils are the most long-lived terrestrial reservoirs for carbon. Terrestrial carbon sequestration takes place in plant biomass (e.g., trees) and living biomass in soil, recalcitrant soil organic C (e.g., hummus), and inorganic carbon in the subsoil. No-tillage, cover crop, and crop diversification provide carbohydrates and nitrogen and allow slow release of nutrients from decaying residues, and thus favors crop growth (Fig. 4.6).

4.4 Adaptation and Mitigation Options in Soil

There are several adaptive soil management practices to cope up with climatic stresses and some of them also help to reduce agricultural GHG emissions. Mitigation and adaptation are seen as separate but complementary targets to be achieved in climate policy. The policy of providing rewards for adaption of the climate-smart management practices could enhance food security and climate change mitigation. Segregating the mitigation and adaptation strategies/options is not scientific as most of the time both are complementary to each other and mitigation followed by adaptation [11]. However, we have placed both approaches separately for easy understanding (Table 4.2).

Table 4.2 Climate-smart adaptation and mitigation of soil practice

Climate-smart soil adaptation practices	Climate-smart soil mitigation practices
Integrated soil fertility management	
(i) Maximizing nutrient supply through organic sources such as compost, green manure, etc. to enhance soil organic carbon builds up	(i) Reducing the input of synthetic nitrogen fertilizers will reduce nitrous oxide emissions
(ii) Enhancing nutrient and water use efficiency through suitable intercropping, mix-cropping, crop rotation, leguminous nitrogen-fixing crops	(ii) Use of controlled release fertilizers, or fertilizers with urease and/or nitrification inhibitors to reduce N_2O emission
(iii) Minimizing inorganic nutrient inputs and reduction of nutrient losses	(iii) Modification of time, method, and rate of manure application to reduce GHG emissions
Biochar application	
(i) The application of biochar promotes soil biological activity and improves nutrient use efficiency (ii) It increases carbon sequestration (iii) The addition of nitrogen fertilizer with biochar could control organic nitrogen mineralization	(i) Reduces greenhouse gas emission by increasing recalcitrant and labile carbon pools ratio
Conservation agriculture	
(i) Conservation agricultural-based practices protect the soil from erosion, reduce evaporation, and promote nutrient conservation	(i) Conservation agriculture reduces carbon dioxide emissions from mechanization due to lower energy requirements for farm operations
(ii) Water and nutrient conservation leads to increasing use efficiency and increased crop production	(ii) It reduces carbon dioxide emissions resulting from different tillage operations
(iii) Conservation agriculture increases the resilience of the soil ecosystem	(iii) Conservation agriculture has a high potential to sequester soil organic carbon. It has been reported that maximum rates of sequestration are achieved in the first 5 to 20 years after the implementation of carbon-enhancing changes in land management. Sequestration rates then decrease until soil organic carbon stocks reach a new equilibrium after 20 to 30 years [16, 12]
Organic production systems	
It does not depend on inorganic fertilizers or pesticides but uses crop rotation, residues, mulch, composts, and green manures to replenish soil carbon	Soil management options in cropping systems include managing fallow periods to increase soil carbon buildup

(continued)

Table 4.2 (continued)

Climate-smart soil adaptation practices	Climate-smart soil mitigation practices
Agroforestry systems	
These systems integrate crops with trees and thus help to restore soil organic carbon and nitrogen	Perennial crops and trees can sequester substantial amounts of C in long-lived stable pools than annuals in the biomass

Source Climate-smart agri source book [5]

4.5 Limitation, Potential, and Possibilities

The major limitation of soil management for CSA is the lack of awareness among farmers, developers, and policymakers. The soils have huge buffering capacity. And higher soil-C emissions and soil-C sequestration both are a silent and gradual long-term processes. So immediate effects of those are not seen and understood by common stakeholders. Therefore, visionary approaches and long-term planning and policies are necessary for soil management in CSA. It is already discussed in previous sections of this chapter that soil has a huge potential to mitigate GHG emissions and C-sequestration. However, the net positive feedback of adaptation, as well as mitigation, could be achieved when the climate-smart practices are framed based on sustainable soil health management. Moreover, increases in soil organic carbon stock can be achieved through policy-driven sound soil and crop management practices.

The recycling crop residues and organics in the soil is one of the best management practices to increase soil organic carbon stock. Minimizing the intensity of tillage that promotes soil aggregation and increases SOC stock is a promising technology in the temperate region but also nowadays becomes popular in the tropics and sub-tropical region. Residue retention along with no-tillage in drylands also has the significant potential to increase SOC stock and crop production. Although an old concept but still valid is the integrated nutrient management in agriculture including horticulture, pasture, rangelands, forestry, and agroforestry. The application of nutrients from combined organic and inorganic sources is one of the best strategies to meet the goal of food security and soil carbon sequestration. Covering the soil surface during fallow periods is another effective option to reduce GHG emissions and enhancing soil-C stocks [29].

Finally, healthy soil is the basis for sustainable production and ecosystem functions. The management practices should aim to increase soil organic matter to create a healthy soil that supports ecosystem functions. The higher inorganic input-intensive production systems need to be converted to integrated production systems, which would further lower the emissions of GHGs. The capacity building program is the need of the hour to develop farmers' knowledge and skill to adopt climate-smart soil management practices to cope up with changing climatic scenarios.

References

1. Baker JM, Ochsner TE, Venterea RT, Griffis TJ (2007) Tillage and soil carbon sequestration—What do we really know? Agric Ecosyst Environ 118(1–4):1–5
2. Bhattacharyya P, Padhy SR, Shahid M, Dash PK, Nayak AK, Gangaiah B (2017) Soil quality index under organic farming. In: Gangaiah B, Kundu A, Abirami K, Swain S, Subramani T, Zamir Ahmed SK (eds), Organic farming in tropical islands of India ICAR-Central Island Agricultural Research Institute, Port Blair, pp 260–267. ISBN : 978-93-85418-30-3
3. Bünemann EK, Bongiorno G, Bai Z, Creamer RE, De Deyn G, de Goede et al (2018) Soil quality–a critical review. Soil Biol Biochem 120:105–125
4. Cerri CEP, Easter M, Paustian K, Killian K, Coleman K, Bernoux M et al (2007) Predicted soil organic carbon stocks and changes in the Brazilian Amazon between 2000 and 2030. Agric Ecosyst Environ 122(1):58–72
5. Climate-smart agriculture source book, Module 4 (2013) Soils and their management for CSA, pp 134. https://www.fao.org/climate-smart-agriculture-sourcebook/en/
6. Corsi S, Friedrich T, Kassam A, Pisante M, Sá JCM (2012) Soil organic carbon accumulation and carbon budget in conservation agriculture: a review of evidence. Integr Crop Manag 16
7. Doran JW, Parkin TB (1994) Defining and assessing soil quality. Defining soil quality for a sustainable environment. In: Doran JW, Coleman DC, Bezdicek DF, Stewart BA (eds.) Defining soil quality for a sustainable environment, vol 35, SSSA Special Publication No. 35, ASA and SSSA, Madison, pp 1–21
8. FAO and ITPS (2015) Status of the world's soil resources (SWSR)-technical summary. Rome, Italy
9. FAO (2013) Climate smart agriculture sourcebook. Rome.
10. Fearnside PM, Barbosa RI (1998) Soil carbon changes from conversion of forest to pasture in Brazilian Amazonia. For Ecol Manage 108(1–2):147–166
11. Inter-governmental Panel on Climate Change (IPCC) (2014) Climate change 2014: impacts, adaptation, and vulnerability. Part A: global and sectoral aspects. Contribution of working group II to the fifth assessment report of the intergovernmental panel on climate change [Field CB, Barros VR, Dokken DJ, Mach KJ, Mastrandrea MD, Bilir TE, Chatterjee M, Ebi KL, Estrada YO, Genova RC, Girma B, Kissel ES, Levy AN, MacCracken S, Mastrandrea PR, White LL (eds)]. Cambridge University Press, Cambridge, UK and New York, NY, USA, p 1132
12. Intergovernmental Panel on Climate Change (IPCC) (2007) Fourth assessment report of the intergovernmental panel on climate change. In: Pachauri RK, Reisinger A (eds) Core Writing Team, IPCC, Geneva, Switzerland, pp 104
13. Karlen DL, Mausbach MJ, Doran JW, Cline RG, Harris RF, Schuman GE (1997) Soil quality: a concept, definition, and framework for evaluation (a guest editorial). Soil Sci Soc Am J 61(1):4–10
14. Karlen DL, Parkin TB, Eash NS (1997) Use of soil quality indicators to evaluate conservation reserve program sites in Iowa. Methods Assess Soil Qual 49:345–355
15. Lal R (1994) Soil erosion research methods. CRC Press
16. Lal R (2004) Soil carbon sequestration impacts on global climate change and food security. Science 304(5677):1623–1627
17. Lal R, Stewart BA (eds) (1995a) Need for long-term experiments in sustainable use of soil resources. In: The soil management: experimental basis for sustainability and environmental quality. Lewis Publisher, Boca Raton, FL, pp 537–545
18. Lal R, Stewart BA (1995b) Managing soils for enhancing and sustaining agricultural production. Soil Manage Exp Basis Sustain Environ Qual 1:1–9
19. Lal R (1997) Soil quality and sustainability. Methods Assess Soil degrad 17–30
20. Lipper L, Thornton P, Campbell BM, Baedeker T, Braimoh A, Bwalya M et al (2014) Climate-smart agriculture for food security. Nat Clim Change 4(12):1068–1072
21. Masto RE, Chhonkar PK, Singh D, Patra AK (2007) Soil quality response to long-term nutrient and crop management on a semi-arid Inceptisol. Agric Ecosyst Environ 118(1–4):130–142

22. Moebius-Clune B (2017) Soil health initiatives of the USDA natural resources conservation service (NRCS) (No. 1958-2017-2580)
23. Noellemeyer E, Frank F, Alvarez C, Morazzo G, Quiroga A (2008) Carbon contents and aggregation related to soil physical and biological properties under a land-use sequence in the semiarid region of central Argentina. Soil Tillage Res 99(2):179–190
24. Ogle SM, Breidt FJ, Easter M, Williams S, Killian K, Paustian K (2010) Scale and uncertainty in modeled soil organic carbon stock changes for US croplands using a process-based model. Glob Change Biol 16(2):810–822
25. Ogle SM, Jay Breidt F, Eve MD, Paustian K (2003) Uncertainty in estimating land use and management impacts on soil organic carbon storage for US agricultural lands between 1982 and 1997. Glob Change Biol 9(11):1521–1542
26. Doube BM, Gupta VVSR (1997) Biological indicators of soil health: synthesis. In: Pankhurst CE, Doube BM, Gupta VVSR (eds) Biological indicators of soil health. CAB International, pp 419–435
27. Paustian K, Lehmann J, Ogle S, Reay D, Robertson GP, Smith P (2016) Climate-smart soils. Nature 532(7597):49–57
28. Pierce FJ, Larson WE (1993) Developing criteria to evaluate sustainable land management. In: Proceedings of the eighth international soil management workshop: utilization of soil survey information for sustainable land use, pp 7–14
29. Poeplau C, Don A (2015) Carbon sequestration in agricultural soils via cultivation of cover crops–a meta-analysis. Agric Ecosyst Environ 200:33–41
30. Qin X, Wang H, Liu C, Li J, Wan Y, Gao Q, Liao Y (2016) Long-term effect of biochar application on yield-scaled greenhouse gas emissions in a rice paddy cropping system: a four-year case study in south China. Sci Total Environ 569:1390–1401
31. Schlesinger WH (1997) An analysis of global change. Biogeochemistry. Academic Press, San Diego, CA, USA
32. Stockmann U, Adams MA, Crawford JW, Field DJ, Henakaarchchi N, Jenkins M et al (2013) The knowns, known unknowns and unknowns of sequestration of soil organic carbon. Agric Ecosyst Environ 164:80–99
33. Tubiello FN, Cóndor-Golec RD, Salvatore M, Piersante A, Federici S, Ferrara A et al (2015) Estimating greenhouse gas emissions in agriculture: a manual to address data requirements for developing countries. Estimating greenhouse gas emissions in agriculture: a manual to address data requirements for developing countries
34. Zach A, Tiessen H, Noellemeyer E (2006) Carbon turnover and carbon-13 natural abundance under land use change in semiarid savanna soils of La Pampa, Argentina. Soil Sci Soc Am J 70(5):1541–1546

Chapter 5
Water Management for Climate-Smart Agriculture

5.1 Water Scenario and Climate Change

The agricultural sector accounts for 70% of global water consumption. Thus, maintaining food security with the scenarios of increasing water scarcity is a major challenge by 2050. Further, in the context of burgeoning population and increasing demand for food supply, agricultural production has to be increased by 60–70% globally, in the coming four to five decades [21]. The consecutive two assessment reports of IPCC AR4 [13] and AR5 [14] consistently emphasize that the continuous increase of GHG emissions above the current level may induce further warming and changes in climatic system at a higher intensity than those in twentieth century. Model-based predictions of the average rise in sea level at the end of the twenty-first century (2090–2099) are variable (0.26–0.59 m). However, it can be noted that the model-based range excludes future rapid dynamical changes in ice flow. The anthropogenic activity-induced rise in sea level will continue to go on, even if the concentration of greenhouse gases are stabilized. All these drivers are directly or indirecting affecting the water/hydrological cycle in the world.

5.2 Climate Change Effects on the Hydrologic Cycle

The climatic system is complex and governed by the interactions among the atmosphere, water bodies, land system, snow dynamics, and ocean [13]. Climate change is closely related to the hydrological cycle and leads to significant changes in regional water systems. Thus, there is a need to adopt adaptation and mitigation measures to climate change. The precipitation pattern, evaporation, evapotranspiration demand of crops, and the occurrence of extreme events would be altered under changing climatic scenarios. For the most efficient management of water resources, a precise and comprehensive assessment of possible future climatic scenarios is a necessity.

© The Editor(s) (if applicable) and The Author(s), under exclusive license to Springer Nature Singapore Pte Ltd. 2020

P. Bhattacharyya et al., *Climate Smart Agriculture*, Green Energy and Technology,
https://doi.org/10.1007/978-981-15-9132-7_5

The projected impacts of climate change on hydrological cycle are presented in tabular form (Table 5.1) below.

Table 5.1 Pieces of evidence and projected impacts of climate change on different components of the hydrological cycle of the world

Components	Projected impact
Precipitation	i. Mean atmospheric water vapor showed increasing trends since the 1980s ii. During the twentieth century, average precipitation was increased in high northern latitudes (1901 to 2005), whereas decreased in 10°S to 30°N, since the 1970s iii. In middle latitude, total precipitation showed a decreasing trend, but the number of events with heavy precipitations was increased iv. In the tropical and sub-tropical regions, the intensity and duration of drought increased and there is a declining trend of soil moisture around the globe v. In many regions of the world, tropical cyclones increased, such as since the 1970s the cyclonic events increased in the North Atlantic region (e.g., North Atlantic)
Snow and glacial ice	i. There is a decrease in snow cover in many regions of the world. Snow cover and permafrost layers have been degraded; the freezing and ice breakup dates have been changed for lakes and rivers ii. It is predicted that the impact of climate change on the hydrologic cycle would be more pronounced in the coming century. With the increasing global temperature, there would be changes in atmospheric vapor pressure; intensity and duration of precipitation; and soil moisture distribution. All of those could lead to changes in the hydrologic cycle [3]
Mean annual precipitation	i. Mean annual precipitation is likely to be more in higher latitudes; in some parts of tropics, it may decrease in sub-tropical lower and middle-latitude areas ii. An increase in extreme precipitation events would lead to more risk of flash floods and water stagnation
Changes in sea level	i. It is projected that increased global temperature will lead to the melting of glaciers and ice sheets, thus causing a rise in sea level
Changes in evapotranspiration	i. There would be changes in potential evapotranspiration due to changes in radiation, humidity, temperature, and wind speed ii. The rise in atmospheric carbon dioxide concentration would also affect the physiological activities of plants iii. The increased global temperature would lead to an increase in potential evaporation also
Soil moisture	i. It is projected that soil moisture would decrease in sub-tropics, Mediterranean conditions, and at higher latitudes ii. However, in Eastern Africa and Central Asia, with increased precipitation, soil moisture may increase

(continued)

Table 5.1 (continued)

Components	Projected impact
River flow and runoff	i. The river flow and runoff could be affected due to changes in precipitation, evaporation, and temperature [13] ii. Temperature increases would play a crucial role in coastal areas due to sea-level rise iii. The seasonality of flow would be affected, with the increased flow in peak period or decreased flow due to the prolonged dry period [15]
Effect on groundwater	i. About 98 percent of global freshwater resources are contributed by groundwater and soil moisture. And the groundwater status is strongly correlated with precipitation [15] ii. Climate change may pose a threat to groundwater recharge due to occurrence of frequent drier periods. In case, if the rainfall would come down below the threshold limit, there would not be any net recharge to groundwater in the near future iii. Salinization of aquifers may take place due to increased evapotranspiration under arid and semi-arid conditions in global warming scenarios

5.3 Water Footprint and Agriculture

5.3.1 Water Footprint

The water footprint is a measurement for the quantity of water utilized for the production of particular products or services. The water footprint is calculated starting from the initial stage of production till the final product is achieved. This concept of water footprint has been gaining importance to determine the final market value of any product. The concept of green water footprint is used to calculate the rainwater used by vegetation and blue water footprint is the water from superficial sources or underground water, which is used for supplemental irrigation under rainfall deficit conditions. Water footprints are basically indicators of agricultural-sustainability. The plant-available water obtained from the soil or directly from rainfall is "green water," whereas the water available from groundwater, lakes, and reservoirs for irrigation and other purposes is called "blue water." A large amount of green and blue water is used for agricultural production, processing, and distribution [6]. The global water footprint (WF) of mankind determined by agricultural production is around 92%. Specifically, the global WF for the agricultural sector is 8360 km^3 yr^{-1}, and the value of gross international virtual water flow in relation to agricultural products is around 2038 km^3 yr^{-1} [9]. So, almost a 24.4% [24.4% = (2038 km^3 yr^{-1}/8360 km^3 yr^{-1}*100] of total WF of agriculture is involved in international trade of agricultural materials.

In order to minimize the increasing extraction pressure on finite water resources, the amount of water used per unit of production (water footprint) should be reduced.

The share of the agricultural sector is 99% in terms of global consumptive (green and blue) water footprint [9]. Further, it has been estimated that the water used for crop production would increase at the rate of 0.7% per annum, to meet the demand for a world population of 9.2 billion, in 2050 [17]. Many parts of the globe are already experiencing acute water scarcity [19, 20].

In this context, the concept of water footprint (WF) acts as an indicator of the rapid measurement of the amount of water consumed per unit of crop production [10]. In nutshell, the green WF quantifies the amount of rainwater used during a crop growth period and blue WF computes the amount of surface and groundwater used for the production. Grey WF is the amount of freshwater that is needed for nutrient assimilation, leaching of pesticides, depending on parameters of water quality standards [10]. The WF as an indicator may be useful to fix the target for water use for the production of a particular commodity, with the objective to minimize the WF and pollution per unit of crop production [11]. Till now, there is no existing standard for WF, as it varies across the region and crops. However, a number of studies have been conducted to benchmark water productivity. The water productivity (a ton of crop production per quantity (m^3) of water used), as calculated for crop production, is actually the reverse of green-blue WF (quantity (m^3) of water used per ton of crop production).

Ercin and Hoekstra [7] developed a possible water footprint scenario for 2050. A number of drivers such as population growth, economic trend, trade scenario, dietary changes, use of bioenergy, and technological factors were taken into consideration (Fig. 5.1).

Their analysis showed that water footprints can drastically vary in future depending on various scenarios and are sensitive with respect to drivers of change (Fig. 5.1):

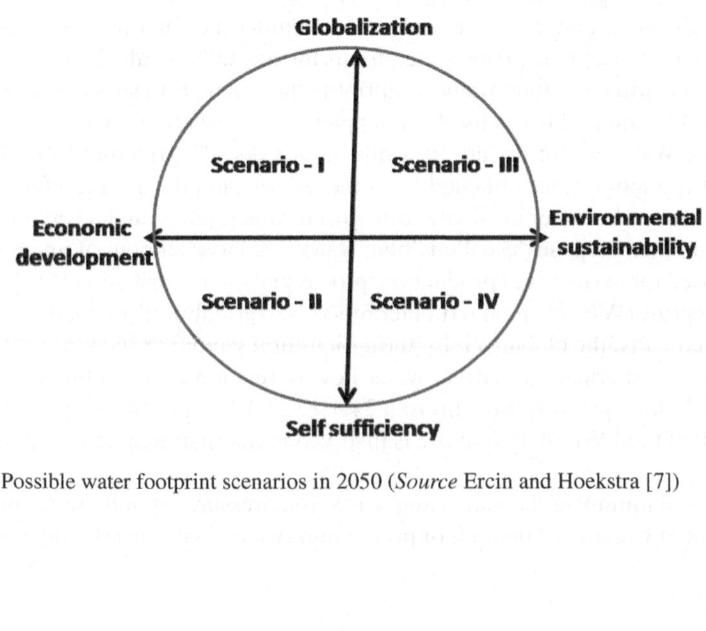

Fig. 5.1 Possible water footprint scenarios in 2050 (*Source* Ercin and Hoekstra [7])

i. **Population growth**: Population size is a major governing factor for WF of production as well as consumption.
ii. **Economic growth**: An increase in income levels may cause a shift to water-intensive commodities. It has been found that high GDP growth promotes more consumption of industrial water.
iii. **Consumer preferences**: Diets rich in dairy and meat-based foods will increase WF, whereas the total water footprint of production drastically decreases by the use of non-dairy and meat products.
iv. **Biofuel use**: A high demand for biofuel will increase the water footprint.
v. **International trade**: A decrease in water footprints may be possible during 2050 by trade liberalization.
vi. **Technology**: Increased water use efficiency by improved technology will reduce WF.
vii. **Policy**: Corrective policies are to be developed for the future to ensure the sustainable use of water resources.

5.3.2 Water Conservation and Integrated Watershed Management

5.3.2.1 Climate-Resilient Watershed Management

Developing climate-resilient watershed is a basic adaptation strategy under changing climatic scenarios. A climate-resilient watershed acts as a storehouse of water and supplies water during non-rainfall periods. A climate-resilient watershed also contributes to control water-induced soil erosion, decrease carbon loss, and ensure the water supply for supplemental irrigation.

A watershed is a hydrological unit in which the runoff drains through a common point in the drainage system. The framework to integrate climate change action plan into watershed management is presented in Table 5.2. Combining efforts from small and marginal farmers in agriculture through the integrated watershed management approach, the restoration of natural resources could be possible in a sustainable manner. Those are also contributing toward mitigation and adaptation to climate change as per the theme of "Sustainable Development Goals (SDGs)." Out of the 70 SDGs along with 169 associated targets, it has been found that 6 SDGs and 16 targets are associated with issues related to watershed management and climate change. The watershed management can successfully be adapted to facilitate successful linkages between up and downstream populations. There are several common approaches to climate change adaptation and mitigation with several watershed management, which are as follows:

(i) Watershed management not only provides environmental benefits but also ensures livelihood, food security, and poverty alleviation benefits along with mitigation to climate change.

Table 5.2 The "Sustainable Development Goal (SDG)" targets to watershed management and climate change adaptation and mitigation

Sustainable development goal	Targets to climate change adaptation and mitigation and watershed management
Ending hunger; food security; improved nutrition; and promoting sustainable agriculture	i. Doubling agricultural productivity ii. Securing income of food producers iii. Ensuring sustainable food production iv. Implementing climate-resilient agricultural practices v. Maintaining genetic diversity
Ensuring sustainable management of water, sanitation for all	i. Improving water quality ii. Implementing water resource management iii. Restore water-based ecosystems iv. Improving water use efficiency
Ensuring sustainable consumption	i. Reducing food loses ii. Sustainable natural resource management
Action to combat climate change impacts	i. Strengthening resilience ii. Adaptive capacity against climatic hazards iii. Integrating measures into policies iv. Building capacity for climate change mitigation v. Impact reduction
Inclusive sustainable development	Representative decision-making for inclusive development at all levels

Source FAO [8]

(ii) This approach offers greater community benefits than individual-centric benefits from adapting climate change interventions even at the household level.

(iii) Watershed management is an integrated approach, which allows groups of stakeholders and organizations to work together to resolve the issues cohesively by maintaining the ecosystem services.

(iv) This approach helps to achieve multiple objectives like strengthening communities to climate change adaptation, decreasing greenhouse gas emissions, conserving the environment, and livelihood security.

(v) However, a sustained joint effort is needed within the framework of integrated watershed management, taking into consideration livelihood security, climate change resilience, and mitigation.

(vi) Systematic integration of climate change adaptation and mitigation aspects into integrated watershed management would not only offer the environmental benefits at the watershed scale but also provide food and livelihood security in addition to climate change mitigation.

The framework to integrate climate change mitigation and adaptation action in the watershed management program includes the themes of agriculture, water resources, energy management, and GHG emission mitigation. The intervention points to agriculture and water resources management include cultivation and incorporation of green manure crops into the soil, addition of soil amendments, avoiding residue burning, and vermicomposting. In water resource management, the interventions are water budgeting, promotion of micro-irrigation, and development of structures for recharging in bore wells. All these interventions could increase the soil moisture conservation, drought resilience, decreased chemical fertilizer use, correcting nutrient deficiencies, increasing soil organic carbon, improving soil health, increased water use efficiency, reducing GHG emission from rice fields, enhancing water use efficiency, and could reduce the bore well drying.

5.4 Integration of the Climate Change Agenda in Watershed Development Programs

It is important to focus on increasing overall resource-use efficiency and introduce technologies for greater water use efficiencies and carbon-neutral life. In this context, the integration of water resource conservation, climate change mitigation, and watershed development should work together by considering some mutual actions and sharing responsibilities (Table 5.3).

5.5 Water Management Strategies under CSA

Climate change is a multidimensional phenomenon, which makes the adaptation to it a complex task [1]. For example, various sectors affected by climate change are closely interlinked (water, agriculture, industries) with each other; there exists a large-scale spatial variation (local, regional, national); and the type of adaptation measures (physical, technological, policy interventions) also varies with the different climatic zones (dry land, mountains, polar and arctic regions). In this section, we will focus on water management in CSA which has direct and indirect impacts on climate change adaptation and mitigation. The water management includes dry land, *rainfed* ecologies, and coastal ecosystems (Table 5.4) [2].

The water management in stress conditions under CSA can be listed under the categories of water allocation, water augmentation, demand-based water management, and water storage (Table 5.5) in water-deficit conditions. The management under water-surplus conditions mostly focused on riverine flood protection and drainage (Table 5.6).

In order to increase the overall water use efficiencies in CSA and irrigation water use efficiency, in particular, there is a need for technological and policy management innovations toward sustainable management of irrigation water (Table 5.7).

Table 5.3 Common aspects and actions need to share to integrate climate change agenda with watershed development activities

Aspect	Action
Watershed strategy for ecosystem management	i. The carrying capacity of any particular watershed is to be maintained sustainably ii. The watershed platform should be used for awareness generation and climate change mitigation action iii. A well-trained watershed community should play a lead role in taking climate-sensitive actions
Proactive policy	i. Watershed program should identify the impacts of climate change on water resources and agriculture and decide well-focused actions for adaptations ii. Water resources should be utilized for increasing the use of efficiencies and capacity building, and specific actions related to climate change mitigation and adaptation iii. The focus should be given on plantation in watersheds and carbon credits involving local communities iv. Adequate facilities should be provided for communicating weather forecasting information to the farmers
Watershed and action plan on climate change	i. Awareness must be created on a carbon-neutral economy in watershed villages ii. Power and energy consumption to be minimized by adopting the best technological options iii. A common program on climate change can be conducted in watershed villages
Social adaptation	Sociocultural on climate change adaptation should be promoted to the communities in watershed villages
Technology	Affordable, accessible, and efficient technology transfer should be done at the community level
Mechanism	Awareness should be generated about clean development mechanisms and carbon financing
Capacity building	To address climate change challenge the communities should be empowered to manage the ecosystems for sustaining their livelihoods

Source Reddy [16]

Table 5.4 Climate-smart water management practices

Parameter	Practices	Outcome
Drought management	In situ soil moisture conservation; mulching; ridge and furrow; raised bed and furrows; broad bed furrow	Increased water productivity
	farm ponds; check dams and tanks	Increased water productivity and cropping intensity
	Conservation agriculture	Resource conservation and sustainability
	Recharging wells; percolation ponds/tanks	Enhanced water productivity
	Contingency crop planning; modified planting techniques; crop residue recycling	Increased productivity and system resilience
Floods; cyclone; excess rainfall conditions	Land shaping in vulnerable coastal tracts	Sustainability of production; increase in water use efficiencies
	Efficient drainage/conveyance system	Increased water productivity and resilience
	Introduction of flood-tolerant crop varieties	Increased water productivity and resilience
	Integrated farming system	Livelihood security; increased income

Source Sikka et al. (2016) [18, 19]

Table 5.5 Water management in water-deficit conditions

Water scarcity condition		
Water allocation	Seasonal weather forecasting and basin level modeling	A suitable combination of hydrological models may be used to estimate water availability, which thereafter should be used for impact assessment of various allocation management options
	Rationing of seasonal water	The aim of rationing is to ensure equitable use among various users, and high water productivity
	Reallocation of water	Resource reallocation will help to cope up with unforeseen situations like water crisis during the drought period; minimize pressure on renewable water supplies; also will help to optimize water use benefits

(continued)

Table 5.5 (continued)

Water scarcity condition		
Water augmentation	Rainwater harvesting	This practice helps to increase rainwater use/entry through the soil surface, groundwater, and plant roots. Rainwater harvesting decreases risks of soil erosion and thus prevents soil and nutrient loss
	Conjunctive use of ground and surface water	The combined utilization of ground and surface water leads to an increase in the resilience of water supply systems. This is particularly effective in regions with high variation in water availability throughout the seasons
	Managing aquifer recharge	In this method, the aquifers are recharged to recover and reuse the water. A managed recharge reduces environmental risks
Management of water efficiency and demand	Managing water efficiency in industries	There is a need for operational and technological improvements to increase industrial water use efficiency. The measures may include leak detection, repairing pipelines, and use of modern technologies like smart systems, timers, smart cleaning, and water monitoring systems. On a policy level, industrial water use efficiency can be increased by providing incentives for water conservation
	Increasing irrigation efficiency	Technologies for increasing irrigation efficiency include using an efficient irrigation method that applies water in a controlled manner based on crop needs Use of wireless sensors along with GPS for site-specific, precise water delivery to meet the need for soil and crop Changing farming practices like the selection of less water requiring crop, conservation tillage, and other in situ soil moisture conservation methods

(continued)

Table 5.5 (continued)

Water scarcity condition		
	Water metering	It is a technique that helps the user to monitor the rate of water consumption and may be combined with pricing charges per unit used
	Reducing loss in water conveyance system	The water loss can be reduced by installing leak detection technique, controlling pressure, maintaining water meters, and adopting all methods that minimize apparent water losses
	Mass campaigns for water conservation	On a large scale, the awareness campaigns should be organized on the social, economic, and environmental utility of water conservation
	Hydrological zoning	In this approach, the land is divided into various zones depending on hydrological properties and specific management practices are adopted for each zone. This approach can protect local water resources from over-abstraction, salinization, pollution, and waterlogging
Water storage	Surface reservoirs	Building structures for surface water storage help to improve water security. The stored water may be utilized for irrigation, domestic, industrial, and power generation purposes
	Multipurpose dams	Multipurpose dams help to combine storage and supply of water for irrigation, domestic, and industrial uses
	Natural wetlands	Wetland ecosystems may be used for climate benefits. Restoration of wetlands will help to restore original hydrology of wetlands in such a way that the benefits of water storage can be harnessed

(continued)

Table 5.5 (continued)

Water scarcity condition		
	Rainwater harvesting	This is *ex situ* practice for water harvesting, where rainwater is collected and stored for multiple purposes. Water harvesting tanks are the most common forms of rainwater storage
Alternative sources	Solar water distillation	Solar energy is used to separate salts and other contaminants from freshwater. Simple solar water distillers may be used for this purpose
	Fog harvesting	This is used to capture water from fog, driven by wind. This is successful in coastal and mountain regions, with naturally high fog
	Inter-basin transfers	In this process, surplus water from the donor basin of a watershed is transferred to the recipient basin with high water scarcity, via pipeline and canal system networks
	Groundwater prospecting	Groundwater prospecting is a promising method to enhance water supply and is thus, an adaptation strategy during water scarcity due to climate change. This can be done by methods like hydrogeological investigations, geological surveys, remote sensing, etc.
	Tube wells; boreholes	As a drought response measure, the freshwater may be extracted from sub-surface or groundwater aquifers, by making boreholes, tube wells, etc.

Source Bertule et al. [4]

Table 5.6 Water management for excess water conditions

Management strategies	Technologies	Description
Riverine flood protection	Creating structural barriers	Permanent constructions are to be made on the path of waterways for containing water against the barrier. Such infrastructure may be in the form of dams, dikes, levees, etc.

(continued)

Table 5.6 (continued)

Management strategies	Technologies	Description
	Optimizing reservoir operations	A number of simulation models are used for this purpose. The algorithms are used to calculate the balance between storage and release of water and thus may be used to device the best strategy to reduce the risk of flood
	Perforated dams	Dams are constructed to control flood in downstream. The spillway is constructed at the level of the riverbed. In case, the water levels reach above the spillway, the dam restricts flow and decreases flood flow
	River restoration	The techniques for restoration are reconnecting rivers with floodplains, the reestablishment of the river's meandering form with no barriers along its banks, and stabilizing soil to decrease sedimentation and soil erosion. This technique is highly effective to manage flood risk
	Lowering glacial lakes	In this process, water from glacial lakes is drained by digging a canal, which drains water from the lake to a surrounding river. Thus, the water level is reduced and the risk of sudden burst events decreases

Bertule et al. [4, 12]

5.6 Limitation, Potential, and Possibilities of Water Management

Two complementary approaches are there for the holistic management of water resources under CSA. These are (a) augmenting the present water reserves and (b) restricting the end-user demand. For coping with climate change, the traditional options are creating new dams; water storage structures and improving water use efficiency are still valid. Moreover, few other recommendations to cope up with climate change issue in relation to water resources are as follows:

(i) Comprehensive planning that includes a strong water resource monitoring program should be figured out. The planning, decision-making, and management should be flexible and easily adaptable, especially in vulnerable areas.

Table 5.7 Management practices for increasing irrigation water use efficiency

Management practice	Outcome
Minimizing the loss of water	The loss of water during conveyance and distribution should be minimized. In this regard, modern technologies like remote sensing and GIS may be useful. Rehabilitation of old water management projects will help to save water
Improving the efficiency of the irrigation system	The efficiency of the sprinkler irrigation system may be improved by optimizing the spacing of sprinkler, right design for variation in pressure, using pressure regulators in hilly and sloping terrain, avoiding irrigation during high winds, and overall good maintenance of irrigation systems The efficiency of the local irrigation system can be improved so that volume of water application is decreased and water productivity is increased. The efficiency may further be increased by using micro-sprayer systems in case of soil with very high infiltration rate, application of irrigation at the right time, and adoption of efficient automation technique
Increasing irrigation water use efficiency	The increase in water use efficiency can be met out through the right irrigation scheduling as per crop demand, proper advisory to the farmers, and selection of best agronomic and soil management practices
Introduction of innovative irrigation methods	In water scarcity areas, the practices like regulated deficit irrigation, sub-surface irrigation, fertigation, and chemigation may be promoted
Policy for water pricing	The policy for water pricing such as volumetric water accounting method may be helpful. Water tariffs and drought surcharges may be promoted. Water pricing policy may affect cropping patterns, water management, and generation of additional revenue that can be used for maintenance of water management projects
Recycling of reclaimed marginal waters	Reclaimed waters may be applied under some guidelines for irrigation to fodder and row crops. Along with supplying water, such water supplies nutrients to the soil and reduces application of chemical fertilizers
Effective participation	Effective participation of private sector, NGOs, public sector, and end users in planning, execution, and decision-making for adoption of water management is required for successful integrated water management

Source Chartzoulakis and Bertaki [5]

(ii) Contingent planning to be developed to meet any kind of climatic uncertainty related to water scarcity.
(iii) A comprehensive evaluation of existing technical approaches for water management should be done in view of possible climate change scenarios.
(iv) A systematic examination is to be done with respect to designs of water harvesting structure, contingency plans, and water-related policies under all the possible climate change scenarios.
(v) Strong linkage between water agencies and scientific institutions is required to facilitate technology innovations.
(vi) Judicious utilization of traditional along with alternative forms of water sources together should be explored to meet the demands and supply scenario, under changing climatic conditions. The alternatives like recycling of reclaimed wastewater, rainwater harvesting, and limited desalination may be looked into.

References

1. Adger WN, Agrawal S, Mirza MMW, Conde C, O'brien KL, Pulhin J, Takahashi K (2007) Assessment of adaptation practices, options, constraints and capacity. In: Parry ML, Canziani OF, Palutikof JP, van der Linden PJ, Hanson CE (eds) Climate change 2007: impacts, adaptation and vulnerability. Contribution of working group II to the fourth assessment report of the intergovernmental panel on climate change. Cambridge University Press, Cambridge, UK
2. Adger WN, Arnell NW, Tompkins EL (2005) Successful adaptation to climate change across scales. Glob Environ Change 15(2):77–86
3. Bates B, Kundzewicz Z, Wu S (2008) Climate change and water. Intergovernmental panel on climate change secretariat. Technical Paper of the Intergovernmental Panel on Climate Change, IPCC Secretariat, Geneva, p 210
4. Bertule M, Appelquist LR, Spensley J, Trærup SLM, Naswa P (2018) Climate change adaptation technologies for water: A practitioner's guide to adaptation technologies for increased water sector resilience. UNEP DTU publication, p 56. ISBN 978-87-90634-07-0
5. Chartzoulakis K, Bertaki M (2015) Sustainable water management in agriculture under climate change. Agric Agric Sci Procedia 4:88–98
6. Döll P, Hoffmann-Dobrev H, Portmann FT, Siebert S, Eicker A, Rodell M, Scanlon BR (2012) Impact of water withdrawals from groundwater and surface water on continental water storage variations. J Geodyn 59:143–156
7. Ercin E, Hoekstra AY (2012) Water footprint scenarios for 2050: A global analysis and case study for Europe. Value Water Res Rep 59:70
8. FAO, Joosten K, Grey S (2017) Integrating climate change adaptation and mitigation into the watershed management approach in Eastern Africa: discussion paper and good practices booklet, pp 66
9. Hoekstra AY, Mekonnen MM (2012) The water footprint of humanity. Proc Natl Acad Sci 109:3232–3237
10. Hoekstra AY, Chapagain AK, Mekonnen MM, Aldaya MM (2011) The water footprint assessment manual: setting the global standard. Routledge Earthscan, London, UK
11. Hoekstra AY (2019) The water footprint of modern consumer society. Routledge, London, UK
12. Hoekstra AY (2013) Wise freshwater allocation: Water footprint caps by river basin, benchmarks by product and fair water footprint shares by community. Value of water research report series. UNESCO-IHE, Delft, The Netherlands

13. IPCC (2007) Summary for policymakers. In climate change 2007: The physical science basis. Contribution of working group I to the fourth assessment report of the intergovernmental panel on climate change, Solomon S, Qin D, Manning M, Chen Z, Marquis M, Averyt KB, Tignor M, Miller HL (eds), Cambridge University Press, Cambridge, UK and New York, USA
14. IPCC (2014) Synthesis report. In: Pachauri RK, Allen MR, Barros VR, Broome J, Cramer W, Christ R et al Climate change 2014: synthesis report. Contribution of working groups I, II and III to the fifth assessment report of the intergovernmental panel on climate change, p 151
15. Kundzewicz ZW, Mata LJ, Arnell NW, Döll P, Kabat P, Jiménez B, Shiklomanov IA (2007) Freshwater resources and their management. Climate change 2007: Impacts, adaptation and vulnerability. Contribution of working group II to the fourth assessment report of the intergovernmental panel on climate change, Parry ML, Canziani OF, Palutikof JP, van der Linden PJ, Hanson CE (eds), Cambridge University Press, Cambridge, UK, pp 173–210
16. Reddy GC (2011) Integration of climate change adaptation strategy with the watershed based sustainable rural livelihood approach, p 10. https://iasc2011.fes.org.in/papers/docs/5000/submission/original/5000.pdf
17. Rosegrant MW, Ringler C, Zhu T (2009) Water for agriculture: maintaining food security under growing scarcity. Ann Rev Environ Resour 34
18. Sikka AK, Islam A, Rao KV (2016) Climate smart land and water management for sustainable agriculture. In: Abstract 2nd World Irrigation Forum (WIF2). 6–8 November 2016. Chiang, Mai, Thailand. W, 3.1.04. pp 1–9
19. Sikka AK, Islam A, Rao KV (2018) Climate-smart land and water management for sustainable agriculture. Irrig Drainage 67(1):72–81
20. Wada Y, Van Beek LPH, Bierkens MF (2011) Modelling global water stress of the recent past: on the relative importance of trends in water demand and climate variability. Hydrol Earth Syst Sci 15(12):3785–3805
21. Zoumides C, Bruggeman A, Hadjikakou M, Zachariadis T (2014) Policy-relevant indicators for semi-arid nations: the water footprint of crop production and supply utilization of Cyprus. Ecol Ind 43:205–214

Chapter 6
Energy Management for Climate-Smart Agriculture

6.1 Energy Scarcity and Climate Change

Worldwide, food and agricultural sectors utilize 30% of total energy resources and contribute approximately 20% of the global emission of greenhouse gases [11]. Maintaining the sustainability of food, and energy sectors are the two biggest challenges before us. The global human population is increasing at an alarming rate and estimated to be 9 billion by the year 2050. To feed this burgeoning population, total food production has to be doubled. The atmospheric level of carbon dioxide (CO_2) gas was 280 ppm in 1750 and it increased to 412 ppm in 2020, which is further increasing at a very high annual rate ~2 ppm per year [15]. Further, there is a possibility of a 25% decrease in production from major grain yielding areas of the world, if the global mean temperature keeps on increasing at the same rate.

The sustainability of the energy sector has a direct impact on food security, under changing global climatic scenario. The production of hydro and thermal energy is further regulated by prevailing climatic conditions. This, in turn, expected to impact the cost of agricultural production and food, by controlling energy prices. The changing climatic parameters affect the various energy generation sources such as hydro, wind, solar, and thermal systems, and climate stressors also impact the performance of the energy systems (Table 6.1) [7]. It has also been reported that the energy use has increased from approximately 11.5 exajoules (EJ) in 1860 to approximately 500 EJ during 2009 [21] and is predicted to increase by 55 percent in 2030, in case the present rate of energy consumption persists [34].

Global reserves of fossil fuels are constant with a fixed depletion period. A majority of the oil stock is concentrated in a few locations of the world, making it further a scarce commodity. There is an urgent need to develop renewable energy sources as a viable alternative to fossil fuel. Moreover, burning fossil fuel also acts as a significant contributor to global greenhouse gas emissions. Managing climate risks while improving energy and resource-use efficiency with the simultaneous increase in ecosystem productivity is essential to achieve food security objectives of climate-smart agriculture, with a simultaneous decrease of greenhouse gas emission.

© The Editor(s) (if applicable) and The Author(s), under exclusive license to Springer Nature Singapore Pte Ltd. 2020

P. Bhattacharyya et al., *Climate Smart Agriculture*, Green Energy and Technology,
https://doi.org/10.1007/978-981-15-9132-7_6

Table 6.1 The effects of changing climatic parameters on energy generation sources and probable future requirement

Climate	Temperature increase	Precipitation variability	Extreme events	Sea-level changes
Hydro energy	Temperature-induced glacier melting; So it is required to strengthen dams/and hydropower stations, to meet higher water capacity requirements of hydropower plants	Increased precipitation may increase erosion in hills and induce silt transportation; Precise hydrological forecasting and efficient de-siltation measures are needed to be developed to cope up with the situation.	The hydropower-infrastructures are to be made more resistant to increased occurrence of extreme climatic events like a cyclone, flash-flood, etc.	….
Wind energy	….	Shifting of power plants may be needed in case of shifting of traditional flood-prone areas	Infrastructures to be designed with a capacity of resisting high wind speed	Shifting of wind power generation-plants from risk-prone sites in case of high sea-level rise
Biomass Energy	Higher warming may require high-temperature tolerant crop species	Precipitation uncertainties may require more efficient irrigation systems. Frequent floods may require building more robust dikes and drainage networks	Timely, real-time meteorological forecasting systems need to be strengthened for crop cutting before the occurrence of any extreme events	Efficient drainage and desalinization technologies to be developed
Solar energy	Efficient cooling systems are to be developed to cope up with rising temperatures	A decrease in rainfall may require developing a cooling system for plants	Solar panels should be designed to resist strong winds, to face the extreme climatic events	Shifting of plants in case of high sea-level rise

(continued)

Table 6.1 (continued)

Climate	Temperature increase	Precipitation variability	Extreme events	Sea-level changes
Thermal energy	Increased temperatures may require more efficient cooling systems	Droughts and floods may require improvements in the robustness of plant stations	Improvements in robustness of plant stations required	Shifting of plants in case of high sea-level rise

6.2 Alternate Sources of Energy and CSA

There are several alternate energy management strategies for controlling greenhouse gas emission to the atmosphere along with the meeting of simultaneous requirement of energy from various sources. In this context, renewable energy may plays a crucial role to mitigate changing climatic patterns while providing multiple co-benefits. Renewable energy may provides a more sustainable energy supply, reduced environmental hazards, fewer human health effects, and better social and environmental security. The possible effects of changes in climatic variables are dependent on the degree of global climate change and the magnitude of these impacts is uncertain. Some of the possible energy management strategies toward climate-smart agriculture are listed below:

i. Replacing a certain portion of synthetic fertilizers with organic manures and biofertilizers.
ii. Promotion of the use of renewable energy sources like solar, wind, biofuels, biomass in agriculture.
iii. Adopting fuel-efficient engines and its efficient maintenance.
iv. Efficient heat management of greenhouses.
v. Insulation of cold storage units of perishable agricultural products.
vi. Introduction and promotion of less-input-demanding crop species.
vii. Improving agriculture value-food-chain and reviewing the existing food storage strategies.

6.3 Biofuel and Bio-Ethanol as Alternative Sources of Sustainable Energy

In terms of energy value, the potential global share of bioenergy along with biofuel has been proposed to be 200 to 400 EJ per annum [18]. The future estimate has shown that biomass has huge scope in terms of meeting energy needs for the future, even to the extent of 1500 EJ per annum [30]. The contribution of biomass sources in terms of power production and industrial purposes was estimated to be 6.4 EJ per annum during 2007. Renewable energy holds promise in terms of becoming one of the largest contributors to electricity generation, just after coal [34]. The power generating industries has huge potential to utilize residues from agriculture and forestry sectors. The biomass material to be used should be low in cost, easily available in large quantity, of high calorific level, and dry. The quality and quantity of material required also vary depending on the thermal and chemical processes used for the processing of biomass. The number of agricultural products and by-products is increasingly being utilized to meet the demand for biomass processing industries. Nonfood crops like poplar, willow trees, and switch grass are best suited for this purpose. Biofuels are increasingly being used in transport and produced economically as "First-generation" biofuels and in the production process crop biomass/residues are used as substrates. Algae, selected oil crops, and jatropha are converted to biodiesel through the thermo-chemical process of trans-esterification of lipids. Many grains and sugar crops are utilized for the production of bio-ethanol by fermentation methods. Second-generation biofuel is produced by the thermo-chemical conversion of lingo-cellulosic materials.

6.4 Alternate Energy Sources in CSA

There are various green energy options available that could be a part of CSA as an inclusive strategy. Some examples of those are listed below:

i. In comparison to energy-producing annual-crops, perennials require lesser fertilizer and other inputs, thus, reduced GHG emission and resulted in higher energy savings. Several research findings indicated that changing production systems from energy to biomass crops (such as willows and poplar) led to high carbon saving and decreased emission of greenhouse gases [14, 28, 32].
ii. Keeping in view of changing climatic scenarios, bioenergy, biodiesel, and biomaterial field is to be included in the cropping calendar to achieve energy security.
iii. Windmills and solar collectors have a wider potential in CSA.
iv. Photovoltaic panels.
v. Biogas production units.
vi. Power generators.
vii. Equipment for bio-oil purification.
viii. Fermentation and distillation facilities.
ix. Efficient equipment for water distribution.

However, in agriculture in general and farming communities in specific, a mix of appropriate energy-efficient technologies are required rather than a single approach to make the system output-efficient and resilient to climatic stresses.

6.5 Adaptation and Mitigation Options and Energy Conservation

Modern agriculture is highly dependent on energy input. So, there is an urgent need to harness the benefits of renewable forms of energy along with the optimum utilization of other non-renewable energy sources. Adaptation measures would help to reduce losses and risks in the energy system. Efficient energy management strategies are to be developed and integrated in order to make the energy sector resilient to climatic changes. Similarly, water management systems that require lesser energy inputs are to be designed which will also lower the carbon footprint.

In general, there are many soil and crop management practices, namely, mixed cropping, intercropping, organic farming, conservation tillage, residue management, mulching, integrated farming systems, agro forestry-based land-use systems, integrated farming system based practices have shown good promises to protect the agro-ecosystems (Fig. 6.1). Such practices also provide provision for food, fuel, and timber, which would foster sustainability and make the energy sector more resilient to climate change (Table 6.2). Specifically, a diversified agricultural production system provides a range of products, by-products, and services and makes the systems more resilient under changing climatic scenarios as compared to mono-cropping. The integrated crop and soil management systems increase soil carbon sequestration and help to combat climate change. Further, adopting an integrated farming system approach with the incorporation of energy-producing plants are promising options toward achieving energy security under changing climatic scenarios [9]. We know that the production of inorganic chemical fertilizer being energy-consuming; replacing inorganic chemical-based fertilizers with organic manures and organic-based production

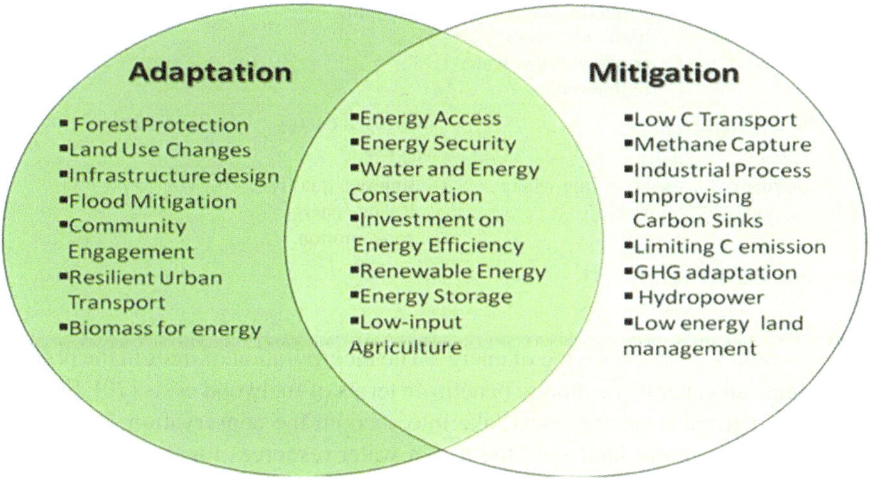

Fig. 6.1 Adaptation and mitigation measures in the energy sector under changing climatic scenarios [27]

Table 6.2 Efficient energy management for crop production

Options	Energy management strategy	Positive and negative effects	
		Positive effect	Negative effect
Conservation agricultural practices	– crop rotation – cover crops – intercropping – grass strips – minimum or zero tillage – mulching – contour cultivation – soil organic carbon build-up – organic farming – integration of the above practices	– Conservation tillage -- Reduce soil erosion, run-off – Conserves water, soil nutrients – Decrease energy use – Increase crop yields – Decrease greenhouse gas emission	Higher workload and time-intensive
Optimum maintenance of mechanical systems	– planned maintenance of mechanized systems, etc	– Reduce energy consumption – Reduce the costs of production – Increase productivity	
Right time input	Increase fertilizer application efficiency	– Higher fertilizer use efficiency – Less fertilizer loss – Saving fertilizer costs	
Increasing water use Efficiency	– water harvesting, recycling – use of precise irrigation methods – high-value cash crops with low water requirements	– Lowering water use – reduce the energy requirement for pumping	
Use of field drying		Decreased energy expense	Pest damage
Use of alternate drying methods	solar drying where possible	– Improve quality – Reduce energy consumption	

Source Smit and Skinner [13, 31]

systems would lead to the saving of energy. The incorporation of trees in the production system can generate economic benefits in terms of fuelwood costs [20]. Sustainable use of natural resources must take into account the conservation of marginal and eroded agricultural lands and the use of water resources for the production of bioenergy crops. In India, sorghum has been identified as the potential bioenergy crop. In this context, "Conservation Agriculture" is a good option which includes minimum tillage and no-tillage practice that reduces the number of farm operations and saves fuel and energy [22]. Reduced farm operations and machinery manufacturing can lead to cut emissions. The other innovative option could be growing of C4 plants which are best suited to drought-prone areas of arid and semi-arid regions, require minimum external inputs, short to medium duration, and grow fast. In order to mitigate climate change, the energy production from fossil fuel has to be reduced and the energy demand of various sectors is to be met out from a variety of energy sources. This will make the energy system more resilient and "climate-smart."

6.6 Climate Change Mitigation Options for the Energy Sector

One of the most commonly utilized bioenergy sources throughout the world is fuel wood. Approximately 15 percent of primary energy especially for household consumption in developing countries comes from this source, and the degree of consumption varies on the national or regional scale. On a global scale, around 70 percent of the energy needs of 34 countries are fulfilled from wood sources, while in 13 countries; fuel wood fulfills around 90 percent of energy needs [33]. Developing countries in Africa, not only faces the challenge of food scarcity but also confronted with issues of bioenergy sources like fuel wood. Growing dependence on fuel wood as an energy source in developing countries is to be restricted to protect the forest, for mitigating global climate change.

Marginal and degraded lands, which are otherwise not suitable for the production of crops may be utilized for the cultivation of bioenergy crops. For this purpose, the best package of irrigation, nutrient management, and post-harvest transport, and processing technologies needs to be developed, keeping in view the economic viability of the production system. This approach would further reduce the emission of greenhouse gases from degraded landscapes. Several bioenergy crops are being grown across the world. Under the Indian condition, sorghum is widely cultivated as a fodder crop and also being increasingly identified as a bioenergy crop. It is grown under the rainfed condition and preferred due to its higher biomass production capacity and water use efficiency. The crop has wider adaptability as it grows under varieties of soil conditions. The growth rate of sorghum is quiet faster even under highly variable temperatures and rainfall conditions of the rainy season. Further expansion of sorghum may be explored in drought-prone arid and semi-arid regions.

In Europe, USA, and Brazil many biofuel crops are grown under rainfed conditions, which are rapeseed, maize, and sugarcane, respectively. To utilize the full potential of bioenergy crops, a wide range of species with diverse genetic variability and hybrids may be grown keeping in view the existing soil and crop production environment [12, 26].

Land-use change by growing a wide range of genetically different species would be more capable of providing a varying range of ecosystem services. As compared to mono-culture, such a system would be more resilient to climatic change and recover faster from disturbances caused by extreme climatic events. The requirement of wood-fuels from the forest sector should be replaced by supply from perennials, or agroforestry would further contribute to mitigating climate change. In India, it has been reported that approximately 24,602 million trees, which are not inside the forest area, contribute to 49 percent of 201 million ton fuel wood, annually, and used in India [25].

There are many ways for decreasing emission of greenhouse gases from the energy sector with the simultaneous provision of energy services [24]. Some of the strategies include,

1. Controlling carbon dioxide emission from the burning of agricultural residues and increasing carbon sink capacity of soil-plant systems [4–6].
2. Reduction in emission of non-carbon dioxide greenhouse gases like methane, nitrous oxides, and hydrofluorocarbons, which can also reduce global warming [23, 29].
3. Replacement of high greenhouse gas energy carriers with low greenhouse gas carriers and renewable energy sources.
4. Utilization of carbon capture and storage method to control global warming.
5. Reducing the requirements of less energy-intensive goods and services [8].

However, the various mitigation measures need to be assessed for their mitigation capacity, cost, and the potential risk factors involved to achieve sustainable development goals (Fig. 6.2).

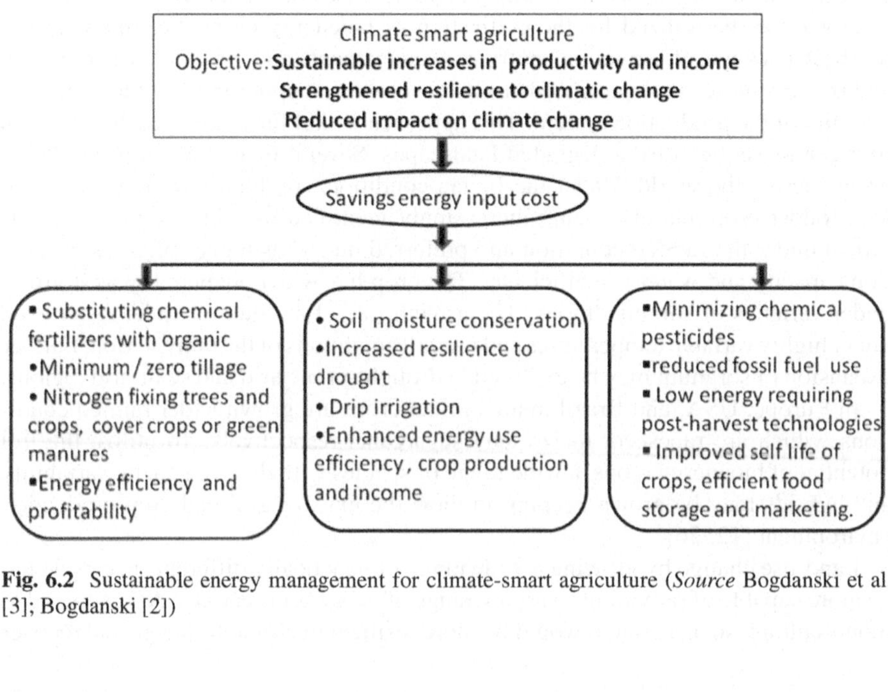

Fig. 6.2 Sustainable energy management for climate-smart agriculture (*Source* Bogdanski et al. [3]; Bogdanski [2])

6.7 Potential and Possibilities of Use of the Renewable Source of Energy in CSA

Under the prevailing changing climatic scenario, more and more utilization of renewable energy may reduce the contribution of the energy sector to climate change and at the same time may play a pivotal role in providing energy security. The demand for renewable energy is increasing day by day as it offers multiple advantages of environmental and health safety, in comparison to fossil fuel.

The possible other benefits using an alternate source of energy in CSA could be listed as, (i) substituting fossil fuels by renewable energy sources can markedly reduce nitrous and sulfur dioxide emission; (ii) possible diversion of waste-land from crop to biofuel production for supporting energy generation; (iii) designing hydropower generating dams may also contribute to preventing the adverse impact of extreme climatic events; (iv) solar panels would also contribute to preventing the adverse impact of extreme climatic events; and (v) large-scale afforestation and forest conservation programs in flood-prone areas could reduce the adverse impacts of climatic change.

6.7.1 Renewable Energy Utilization Potential in Agri-Food Cycle: Global Scenario

Ensuring the availability of environment-friendly and economic energy sources is a key component for energy and climate-smart agriculture. Directly, renewable energy can be used in various steps of the agricultural food production system and it could also be incorporated in activities that are otherwise carried out by using traditional energy sources. Therefore, keeping in view the changing climatic scenario, the maximum utilization of renewable energy sources in agriculture should be encouraged to minimize fossil fuel emission (Fig. 6.3).

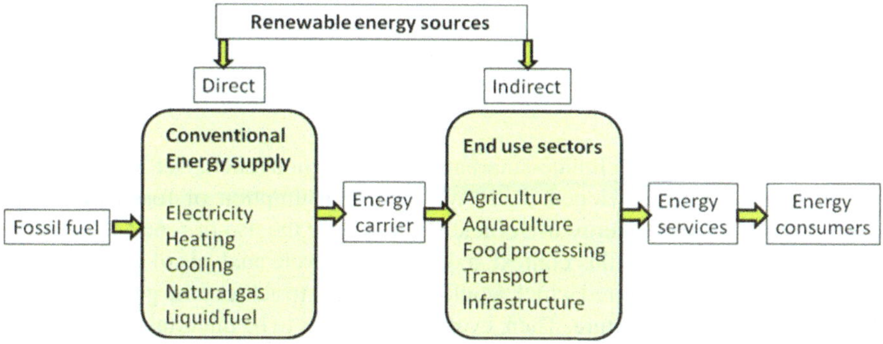

Fig. 6.3 Renewable energy services across various sectors IPCC [16, 17]

Post-harvest storage and processing are some of the major energy-consuming activities in the agri-food chain, especially in developing countries. Therefore, the best possibilities for the use of renewable energy are to be explored to meet this demand. Among the renewable energy sources, solar as well as biomass energy have a great role to play for storage purposes of agricultural produce. Storing of spice crops is a major issue in Sri Lanka and wood biomass is increasingly being used in this sector. Big opportunities exist to generate renewable energy from solar, hydro, and wind sources by best energy production technologies, which can be utilized for mechanized post-harvest operations.

As a primary step to mitigate climate change, the use of fossil fuel has to be replaced by renewable energy sources. Estimates have shown that with the establishment of solar and wind energy production technologies, in around 1.5% of land presently under agriculture, fossil fuel replacement could be achieved with the least impact on the agricultural sector [1]. Currently, 13 percent of world energy requirement is met out from renewable energy sources. Projections have shown that by 2050, this share of renewable energy to meet global energy demand may increase up to 70 percent [16, 17]. In the future, with the advancement of renewable energy-producing technologies, the production cost is expected to decline.

Some waste products are generated in the agri-food processing chain which can be used as biomass resources for energy generation. There is scope for diversification of markets with further expansion of bioenergy production technologies in agriculture. For example, maize and wheat by-products have already been used as feedstocks in biofuel industries. Corn and wheat, for example, could be sold for milling, animal feed or biofuel feedstock. Such diversification would also lead to employment generation in developing countries. Several case studies carried out in 12 countries including Africa, Asia, and Latin America confirmed sustainable use of bioenergy in small farm-based projects with a positive impact on local food security [10]. In Mauritius, sugar cane bagasse is used for the generation of electricity supply [19]. Macro- and micro-algae are aquatic plants with large potential to sequester carbon. The oil yield from many micro-algae species can be higher than vegetable oil. The biomass crops may be utilized in an integrated manner for food as well as energy production with best management practices.

6.7.2 Future Need

To meet the objectives of climate-smart agriculture, a more energy-efficient production system is to be developed, with minimum consumption of fossil fuel. This would further enhance energy use efficiency; promote the use of renewable energy in production systems. In this context, a detailed life cycle analysis of every energy-based intervention is required which would further help to achieve the goal of sustainable climate-smart agriculture. Then, every measure has to be undertaken to enhance energy efficiencies at all stages of production systems and the energy requirement of various sector need to be met out from a diverse range of energy sources.

References

1. Bardi U (2007) Solar power agriculture: a new paradigm for energy production. In: 2004 new and renewable energy technologies for sustainable development, pp 235–247. https://www.spiritviewranch.com/pdf/Christoph/archive/Solar%20Power%20Agriculture.pdf
2. Bogdanski A (2012) Integrated food–energy systems for climate-smart agriculture. Agric Food Secur 1(1):1–10. https://www.agricultureandfoodsecurity.com/content/1/1/9/abstract
3. Bogdanski A, Dubois O, Jamieson C, Krell R (2011) Making integrated food-energy systems work for people and climate: an overview. Food and Agriculture Organization of the United Nations (FAO). https://www.fao.org/docrep/013/i2044e/i2044e.pdf
4. Bond TC, Sun H (2005) Can reducing black carbon emissions counteract global warming? Environ Sci Technol 39(16):5921–5926
5. Bhattacharyya P, Dash PK, Swain CK, Nayak AK, Chatterjee D, Padhy SR et al (2018) Carbon dynamics in soil-plant-environment system on climate change perspective: Special reference to rice. In: Advances in crop environment interaction, pp 3–23. Springer, Singapore
6. Alvarenga P, Carneiro JP, Fangueiro D, Cordovil CM, Bernal MP (2020) Managing organic amendments in agroecosystems to enhance soil carbon storage and mitigate climate change. In: Climate change and soil interactions, pp 89–141. Elsevier
7. Stuart C (2017) Energy and climate change adaptation in developing countries. Published by European union energy initiative. Partnership dialogue facility (EUEI PDF), p 60
8. Dietz T, Gardner GT, Gilligan J, Stern PC, Vandenbergh MP (2009) Household actions can provide a behavioral wedge to rapidly reduce US carbon emissions. Proc Natl Acad Sci 106(44):18452–18456
9. Dixon JA, Gibbon DP, Gulliver A (2001) Farming systems and poverty: improving farmers' livelihoods in a changing world. Food and Agriculture Organization
10. FAO (2009) Small scale bioenergy initiatives—brief description and preliminary lessons on livelihood impacts from case studies in Asia, Latin America and Africa. Policy innovation systems for clean energy security, practical action, and FAO, environment and natural resources management working paper 31. Food and Agriculture Organization of the United Nations, Rome, p 149. https://www.fao.org/docrep/011/aj991e/aj991e00
11. FAO (2011) Food and agriculture organization of the United Nations: energy-smart food for people and climate. Issue Paper, Rome
12. Habyarimana E, Laureti D, De Ninno M, Lorenzoni C (2004) Performances of biomass sorghum [*Sorghum bicolor* (L.) Moench] under different water regimes in Mediterranean region. Ind Crops Prod 20(1):23–28
13. Heumesser C (2015) Climate smart agriculture indicators and index, world bank. https://www.asti.cgiar.org/sites/default/files/pdf/s-t-partnerships/CSA-results-indicators
14. Hillier J, Whittaker C, Dailey G, Aylott M, Casella E, Richter GM et al (2009) Greenhouse gas emissions from four bioenergy crops in England and Wales: integrating spatial estimates of yield and soil carbon balance in life cycle analyses. Gcb Bioenergy 1(4):267–281
15. IPCC (2007) Intergovernmental panel on climate change: synthesis report. In contribution of working groups I, II and III to the fourth assessment report of the intergovernmental panel on climate change. edited by core writing team, Pachauri RK, Resinger A (eds). Geneva, p 104
16. IPCC (2011a) Special report on renewable energy and climate change mitigation, chapter 8, integration of renewable energy into present and future energy systems, working group III, intergovernmental panel on climate change. https://srren.ipcc-wg3.de/report/IPCC_SRREN_Ch08
17. IPCC (2011b) Special report on renewable energy and climate change mitigation, working group III, intergovernmental panel on climate change. https://srren.ipcc-wg3.de/
18. Junginger HM, Faaij APC, Rosillo-Calle F, Woods J (2006) The growing role of biofuels-opportunities, challenges and pitfalls. Int Sugar J 108(1295):618–629
19. Karekezi S, Kithyoma W (2006) Bioenergy and agriculture: promises and challenges. Focus 14
20. Kürsten E (2000) Fuelwood production in agroforestry systems for sustainable land use and CO_2-mitigation. Ecol Eng 16:69–72

21. Lal R (2010) Managing soils for a warming earth in a food-insecure and energy-starved world. J Plant Nutr Soil Sci 173(1):4–15
22. Lal R (2005) Forest soils and carbon sequestration. For Ecol Manage 220(1–3):242–258
23. Moomaw WR, Moreira JR, Blok K, Greene D, Gregory K, Jaszay T et al (2001) Technological and economic potential of greenhouse gas emissions reduction. Technological and economic potential of greenhouse gas emissions reduction. In: Climate change 2001: mitigation of climate change. Contribution of working group III to the third assessment report of the intergovernmental panel on climate change. Cambridge University Press, pp 167–299
24. Pacala S, Socolow R (2004) Stabilization wedges: solving the climate problem for the next 50 years with current technologies. Science 305(5686):968–972
25. Rai SN, Chakrabarti SK (2001) Demand and supply of fuelwood and timber in India. Indian Forester 127(3):263–279
26. Rattunde HFW, Weltzien E, Diallo B, Diallo AG, Sidibe M, Touré AO et al (2013) Yield of photoperiod-sensitive sorghum hybrids based on guinea-race germplasm under farmers' field conditions in Mali. Crop Sci 53(6):2454–2461
27. Winkelman S, Nichol E, Harford D (2017) Taking action on green resilience: climate change adaptation and mitigation synergies
28. Schmer MR, Vogel KP, Mitchell RB, Perrin RK (2008) Net energy of cellulosic ethanol from switchgrass. Proc Natl Acad Sci 105(2):464–469
29. Sims RE, Schock RN, Adegbululgbe A, Fenhann J, Konstantinaviciute I, Moomaw W et al (2007) Energy supply climate change 2007: Mitigation. Contribution of working group III to the fourth assessment report of the intergovernmental panel on climate change. Cambridge University Press, Cambridge, United Kingdom and New York, NY, USA
30. Smeets EM, Faaij AP, Lewandowski IM, Turkenburg WC (2007) A bottom-up assessment and review of global bio-energy potentials to 2050. Prog Energy Combust Sci 33(1):56–106
31. Smit B, Skinner MW (2002) Adaptation options in agriculture to climate change: a typology. Mitig Adapt Strat Glob Change 7(1):85–114
32. Stephenson AL, Dupree P, Scott SA, Dennis JS (2010) The environmental and economic sustainability of potential bioethanol from willow in the UK. Biores Technol 101(24):9612–9623
33. Trossero MA (2002) Wood energy: the way ahead. Unasylva 53(4):3–12
34. Umbach F (2010) Global energy security and the implications for the EU. Energy Policy 38(3):1229–1240

Chapter 7
Crop Management for Climate-Smart Agriculture

7.1 Introduction

Agriculture is vulnerable to climate change consequences (erratic rainfall, drought, floods, cyclones, heatwaves, fluctuation of monsoon, cold wave, hailstorm, etc.) not only for its high dependence on weather variability but also it is practiced mostly (>70%) by small and marginal (resource-poor) farmers globally, having the low coping capacity to stress.

Various crop management practices have been followed and recommended in sustainable agricultural goals, and many of them are well suited for CSA also. The sustainable agricultural development and CSA approaches are not different but complementary to each other. We can say CSA is managing the sustainable agricultural development under the climate change perspective. Crop management helps in maintaining higher agricultural production with increasing input use efficiencies, lowering damage to natural resources, sustaining soil health, and enhancing ecosystem services. Conserving the genetic diversity of plants (crop), growing contrasting varieties in association with a wide range of plant species in rotations/sequences are an important aspect of climate-smart crop management. Efficient water management with integrated nutrients (fertilizers, manures, soil amendments, crop residues) and pest management (insects, diseases, weeds) are an integral part of CSA. The crop is an element of a broader agro-ecosystem in which soil, biodiversity, and ecosystem services are other key elements. As CSA always talks about system approach and environment sustainability, so the crop management practices under CSA must be discussed on a landscape scale. Therefore, we should promote crop management that provides positive feedbacks to sustainable food production and negative feedback to climate change [20, 23]. Practices must rely on ecological processes, generate ecosystem services, increases the resilience and adaptive capacity of agricultural landscapes. There are many proven crop management practices that exist and are implemented worldwide in the last 6–7 decades having multiple benefits

© The Editor(s) (if applicable) and The Author(s), under exclusive license to Springer Nature Singapore Pte Ltd. 2020

P. Bhattacharyya et al., *Climate Smart Agriculture*, Green Energy and Technology,
https://doi.org/10.1007/978-981-15-9132-7_7

for food security and environmental health, and those must be integrated first with CSA to make it more resilient and better adapted by farming communities.

In this chapter, we will focus on the importance of conservation of genetic diversity of crops; efficient crop management strategies and their adaptation and mitigation potential to climate change. Inter-linkage of management practices and few case studies with their limitations and opportunities would be also discussed for easy understanding.

7.2 Conservation of Crop-Genetic Diversities

The process and structure of ecosystems that provide valuable ecosystem services are maintained through balanced biodiversity (diversity of flora and fauna). Maintaining biodiversity is not only essential for agriculture and crop production but necessary for satisfying the different needs of the community as a whole. It is the key regulator of the functioning of agro-ecosystems. On one hand, it enhances the resilience of the agro-ecosystem and also reduces the vulnerability of farmers to climate change risk. Conservation of belowground flora and fauna are equally important from an agricultural point of view as the soil is the primary medium of crop production. Additionally, balanced biodiversity provides a lot of intangible ecosystem services that are beneficial for farmers and sustainable food production. An ecosystem approach to conserve biodiversity (crop, animal, and microorganism) is effective as it would also help to maintain associated plant/animal diversity along with selected/targeted crops.

7.2.1 Importance of Maintaining Crop Genetic Diversity on Climate Change Perspective

Targeting the higher yield of few specific crops for food security at one point of time may force to reduce the maintenance of higher genetic resources. But conserving and maintaining genetic resources have become important when climate change occurs and some biotic stresses prevail. Different genotypes have separate environmental niches and perform differently at diverse climatic conditions [40, 43]. The diversity even exists within the same crop species that allows them to cultivate across different regions under diverse climatic conditions. These potential variations and performance of germplasm in challenging environmental conditions could be exploited to develop tolerant cultivars to stress (abiotic/biotic). So crop genetic resources must be conserved for the future. The crop genetic resources have been continuing to play a critical role in building resilience in agriculture by providing suitable varieties and breeding stocks in changing climatic situations. To ensure food security for burgeoning human population, agriculture has to be adapted to the impacts of climate change. In this context, agriculture highly relies on the genetic resources of crops and animals for developing climate-stress-tolerant varieties/breeds. Intelligent use of crop genetic diversities is the key to crop improvement for getting higher yields, better nutritional quality, reducing the production cycle, and building system resilience. A wide range of genetic diversities of related species of major cultivated crop varieties in the world is the treasures for plant-breeder. They are incessantly trying to exploit their (plat germplasm) genetic potential for developing the productive and stress-tolerant cultivars.

The land and environmental degradation along with climate change are putting threats to the genetic diversity that are being lost continuously at an alarming rate. Highly preferred consumer demand, loss of natural habitat, and commercialization of a few specific species for economic benefits cause rapid erosion of traditional varieties and breeds. Current estimates show that only five cereals namely, rice, wheat, millet, maize, and sorghum contribute 60% of the energy demands of humans globally. More so, only 30 crop species contribute 90% of the food and energy demand of humans worldwide [16, 46]. The number of intensively cultivated crop species provides the humans' protein needs globally even much lesser than that of cereals. However, there are huge genetic diversities that exists even in cultivated crop species. For example, more than 150,000 thousand distinct rice (*Oryza sativa*) germplasm has an immense potential as the donor for developing climate-stress-tolerant varieties [42]. Similarly, the "farming communities in Andes" alone cultivate around 175 potato varieties [15]. Overall, there is around 60 thousand plant germplasm (wild relatives of cultivated crop varieties or else) discovered which can effectively act as gene donors for future plant breeding.

Around 80% of terrestrial biotic (flora and fauna) diversities are found in forests. The evolution of forest species, succession, and their stability primarily depend on their coping capacities to adverse climatic conditions and adaptive potential to adverse conditions. The forest ecosystems have the huge potential to mitigate climate change and GHG emissions along with moderate negative consequence (risk) of climate change vagaries [19]. However, studies and regulations for conservation on forest-genetic diversity are still in the baby stage. For example, the world's forest is the home of around 80 thousand tree species, but <1% of them are still studied for assessing their potential. Moreover, the anthropogenic pressure on this rich forest-biodiversity is immense causing unprecedented deforestation, unsustainable utilization of forest-products, misuse of timbers and forest products, etc. In this connection, just have a look at the microbial diversities that exist on earth and their utilization pattern as they are well connected with agriculture and are necessary for maintaining the plant genetic resources on climate change perspective. The most numerous species found on Earth are micro-organisms and invertebrates. Those include bacteria, fungi, actinomycetes (microorganisms), insects, arachnids, and earthworms (invertebrates). They are contributing immensely to ecosystem services, pest control, nutrient mobilization on soil, nitrogen fixation, organic carbon decomposition, maintaining soil health, and thereby are essential components of agriculture and forest ecosystems. They have also potential use in food processing and preservation industries [18].

7.2.2 Interrelationship between Crop Genetic Diversity and Climate Change

Crop genetic resources related to food and agriculture and climate change are interlinked. Climate change threatens the survival and performance of plant genetic resources; while conserving and utilization of diverse crop-germplasm open the windows of coping with climate change challenges. Climate change severely impacted the structural and functional diversities of crop species. It affects the community composition, distribution, population sizes (structural diversities) in one hand, and the way of responses to biological processes, behavior, and timing of biological events (functional diversities) on the other hand. Climate change-induced change in plant physiology influenced the pathways of enzymatic reaction and composition of plant products. Climatic aberration also causes asynchrony between the abundance of pollinators and flowering time and unprecedented conditions for parasites and pests. Many a time, epidemiology of many crops are also distracted. Climate change can alter the ecosystem dynamics having consequences on crop genetic diversities [32, 53].

Now the other way, the immense potential of plant genetic resources can be effectively used for climate change mitigation through the development of stress-tolerant varieties and imparted intrinsic tolerance to plant systems. The major traits present in diverse plant germplasm that could be utilized for developing stress-tolerant cultivars required to climate change adaptation are, fire resistance and tolerance for forest trees; ability to withstand droughts and high temperatures; and resistance to insects and diseases; higher nutrient and water-extracting capacity from soils; tolerance to submergence and salinity; and having anaerobic seed-germination capacity. However, for exploiting the genetic potential of plant genetic resources for food and agriculture in climate change scenarios, we have to move geographically, and need to use scientific, traditional, or indigenous knowledge supported by transparent international cooperation [12, 24].

7.2.3 Method of Conservation of Crop Genetic Resources

The crop genetic material can be conserved both in-situ and ex-situ conditions. The genetic materials are grown in a natural environment where the species have evolved in the in-situ conservation process. It is a static, labor-intensive, costly way of conservation that needs a lot of physical space but capable of maintaining the genetic-integrity of the material during the time of collection. Botanical gardens, field-gene banks, weed-park, and zoos are examples of in-situ conservation. However, many a time, it is difficult to conserve the huge genetical material in-situ or on-farm (both cost and space point of view) conditions then the alternative effective option is ex-situ conservation. In this method, the material is collected from different potential zones and then stored them in gene-bank with proper cataloging. It facilitates easy accessibility of germplasm with the required information. However, the conserving condition (desired temperature and moisture) must be maintained in the gene-bank with an uninterrupted power supply. The conserved materials in ex-situ gene banks also need periodic rejuvenation in field-condition by maintaining their genetic-integrity. Moreover, the materials in gene-bank need to be stored in duplicate at different locations of the world with reasonable insurance coverage. Properly characterized genetic material in gene-bank are the excellent material for plant breeding in general and abiotic-stress tolerance breeding in particular in changing climatic conditions. Roughly about 7.4 million crop genetic materials are conserved in ~1750 gene banks across the world [16, 50, 66, 72].

7.3 Management of Crop Production in CSA

Climate change affects crop production both negatively and positively. It has significant impacts on the food security and livelihoods of farmers. The impact of climate change on crop production has already been discussed in Chap. 2 in this book. So in this chapter, we will study the climate-smart crop production management that provides farmers to adapt to, and also mitigate, climate change. The climate-smart crop production strategies should fulfill the three basic criteria of CSA, i.e., increase in yield on a sustainable basis, enhance the system resilience through higher adaptive capacity, and mitigate GHG emissions or climate change. The practices include the introduction of stress-tolerant crop variety, change in cropping system on the environment-friendly and economic manner, and managing input resources (water, soil, fertilizers, pesticides, etc.) in an effective way to reduce losses and emissions (GHG). Judicious crop management strategies also have good potential to reduce GHG emissions. Few examples are real-time nitrogen management, avoiding flooding in the rice-paddy field, using sulfur-rich soil amendment to rice-soils, carbon-sequestration through the introduction of agroforestry and silvopasture in wastelands, etc. The sustainable crop production practices already followed by farmers with traditional knowledge have a greater significance to CSA as those take care of local ecosystem balance. The integrated farming system having both crop and livestock components with agroforestry is "climate smart." It provides farmers' diverse source of income, imparted resilience to them, and also having mitigating potential via carbon sequestration.

At this juncture, it is very important to know that CSA (in terms of climate change adaptation and mitigation in agriculture) and "sustainable crop production" are not different rather complementary to each other. The underlying principles of both are nearly the same. The CSA looks at the "sustainable crop production" through the lens of "climate-change." Both talking about the enhancement of sustainable crop (food, fodder, and fuel) productivity, and agro-ecosystems based adaptation [3]. Further, CSA adds mitigation to climate change with food security and resilience. However, many of the sustainable crop production strategies also have the mitigation component in-built with that (e.g., integrated farming systems). Those are perfectly matched with the principles of CSA. From a broad perspective, both are working for the same goal. The goal is to ensure the supply of enough food (safe and nutritious) to all (human) present and future generations.

In this context, diversified cropping system (cereals + legume-pulses + oilseeds + vegetables + spices + fruits, etc.) is the essence of nutritional security of the human population which is increasingly critical along with food security in near future. A wonderful example of CSA is the incorporation of leguminous-pulse in the farming system which is an input-saving resource-conserving technology (fix atmospheric nitrogen), helps in the reduction of nitrogenous-fertilizer use, reduces nitrous oxide (GHG) emission and also inhibits soil pathogens infestation [17]. The CSA approaches acknowledge the threats of climate change in the present as well as future crop production and food security. It also talks about building the resilience of the system to increase the coping capacity of farmers under climate change vagaries. So, ecosystem management on a long-term basis is essential with crop-agronomy to build-up the system resilience. Moreover, for secure crop production in the future, we have to mitigate and adapt to climate change. So, crop management under CSA should follow the ecosystem approach and climate change mitigation in mind.

7.4 Approaches to Productive Crop Management Practices Under CSA

Different productive (high yield potential) crop management practices in agriculture contribute to climate change adaptation or mitigation or both (Table 7.1). In this section, we mainly focus on those practices that address both adaptation and mitigation to climate change. The practices should have high yield potential or high system yield. The typical mitigation and adaptation approaches would be discussed in detail in Chap. 9 in this book. Here, the objective is to introduce the technicality of the management practices in light of CSA. The conservation agriculture and ecosystem-based approaches are typically suited under CSA that addresses the resilience systematically. Of course "integrated farming system," "agroforestry," and "participatory watershed management" are relevant in CSA when addressing the issue of GHG emission mitigation. In the following sections of this chapter, the impact of relevant crop management systems on productivity, resilience, and mitigation would be elaborated with location-specific case studies.

7.4.1 Conservation Agriculture

Reduced tillage, residue retention, and crop diversification are the three key features of conservation agriculture (CA). It is "climate-smart." It ensures sustainable crop yield by providing adequate nutrients and water and facilitates the use of high yielding varieties. At the same time, putting residues to the field (organic matter/carbon addition) continuously improves the soil carbon and nutrient status gradually which saves fertilizer use and thereby reduces GHG emissions. "Reduced tillage" operations help in soil compaction, restores soil health (biological and physical), reduce energy use, and ultimately improves the resilience of the system and mitigate GHG (CO_2, N_2O) emissions. Moreover, the CA helps to reduce the degradation of natural resources (soil, water, energy) and sustain ecological balance by favoring natural biological processes. Sufficient organic carbon (coming from crop-residues) and relatively undisturbed soil environment are congenial for soil-fauna like millipedes, earthworms, mites, and other beneficial microorganisms. Water holding capacity of soil enhanced due to an increase in water-filled porosity (less breaking of soil aggregates due to reduced tillage) and soil-organic-carbon build-up which imparts resilience to the system under drought or moisture stress condition. Moreover, stable soil aggregates and vertical macro-pores created by earth-worms maintain optimum water infiltration rate and provide adequate drainage to excess water that makes the system resilient and adaptive to flooding and erosion. The CA not only reduces GHG emissions but also helps in sequestering carbon in soil and mitigates climate change. Crop diversification with deep-rooted crops or legumes in CA could break subsoil compactions and provide additional nitrogen to soil and plant [5, 6, 15, 17].

Therefore, the ecosystem-services and resilience provided by CA could be listed as, soil moisture-retention; soil-fertility maintenance; sustainable yield increase; reducing yield-variability; enhanced production per area, and ensured food security [54]. The CA could build-up organic carbon status of its soil to become their fertilizer but it takes time. The beneficial effect does not get overnight, it requires 2–4 years "pay-back time" in terms of fertility and carbon build-up in the soil to become more satisfactory. Improved soil health and nutrient-cycling, better soil physical processes and carbon build-up in CA would improve the crop as well as system resilience to adapt to climate change.

In this paragraph, we will discuss some case studies of the impact of CA as a whole and component of CA on crop productivity, resilience, and mitigation aspects. For example, the reduced tillage combined in maize (cultivar: Nilesh)-cowpea intercropping enhanced the system productivity by 200% over traditional farmer's practice (conventional tillage) of maize cropping in India and achieved 300% additional net benefits. The soil quality as an indicator of regulatory ecosystem services is also highest under CA. In that way, the CA simultaneously increased system productivity, crop-diversity, and improve soil quality and make the practices "climate-smart" [56]. Similarly, another CA practice in rice-green gram cropping system in eastern India having the components of dry direct seeding of rice and *dhaincha* in paired row cropping (seed drill sowing); green manuring/brown manuring; and real-time nitrogen management (by CLCC) enhanced the system yield by 9.6% over farmers practice, saved 32% of energy, and reduced 5% of methane emission [8]. Another study from India revealed that the no-tillage and residue management in the rice-wheat cropping system in Indo-Gangetic delta sustained the higher system yield with reduced GHG emissions [51] (Fig. 7.1).

7.4.2 Site-Specific Agro-Ecosystem based Approaches with Stress-Tolerant Cultivars

Agro-ecosystem based crop management practices are primarily based on local resource availability, ecological niche, and vulnerability of farmers. Traditional and stress-tolerant cultivars with the low-input requirement are the key criteria for choosing the crop varieties to be grown and cropping system to be followed in this approach. It gives more importance to ecosystem-resilience and the adaptive capacity of farmers (Fig. 7.2). The potential abiotic stress-tolerant varieties of rice, wheat, and maize (Table 7.2) are found promising all over the world with modified agronomic practices.

Apart from stress-tolerant cultivars, ecology specific agronomic practices are also important. In this connection, in-situ moisture conservation, direct-seeded rice, and site-specific nutrient management are practices in *rainfed* upland ecology for rice. Whereas, proper drainage networking, with the deep placement of urea briquette (nitrogen management) and post-flood applicant NPK (spray), are recommended for flood-prone areas. Frequent irrigation with good quality water, organic manure application in sufficient quantity with drainage provision for removing saline water are the necessary practices that need to be followed in saline zones.

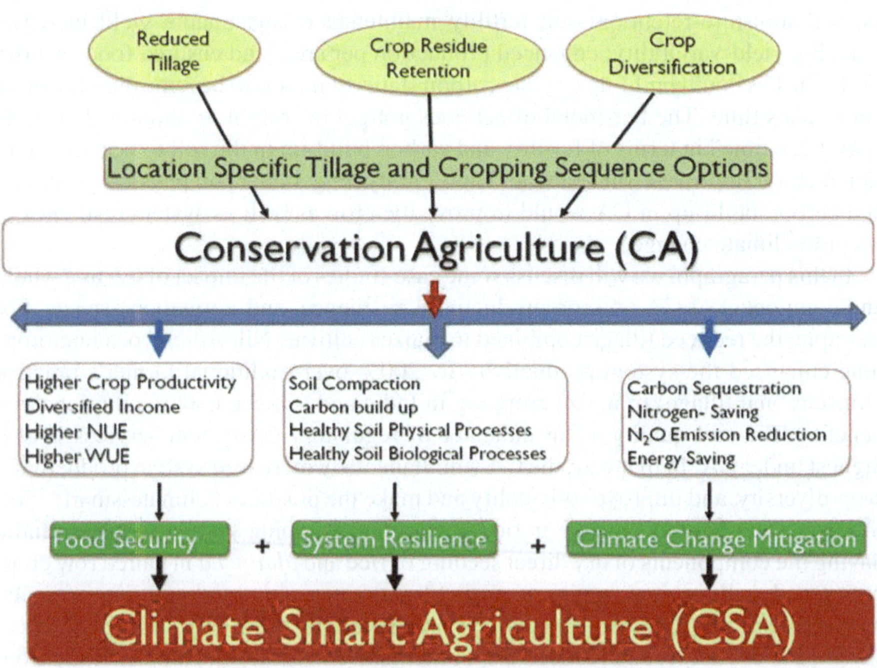

Fig. 7.1 Conservation agriculture (CA) is a climate-smart technology

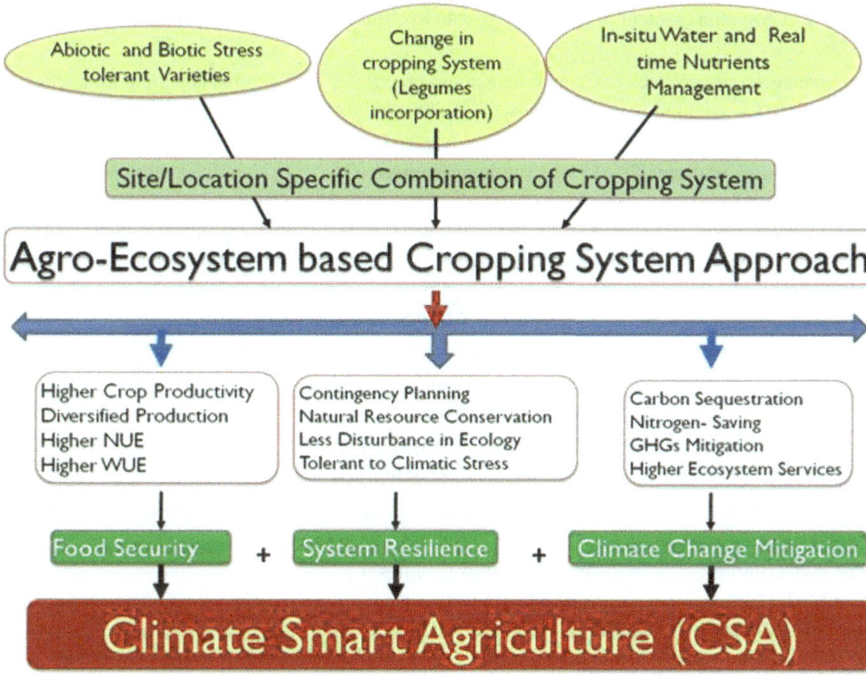

Fig. 7.2 Agro-ecosystem based cropping system approach is Climate-Smart

For example, ecology-based rice crop production management in light of climate-resilient as well as CSA are presented in Table 7.3 for better understanding.

7.4.3 Integrated Farming System

Integrated farming system (IFS) refers to the growing and maintaining of compatible crops (field, horticulture, grasses, etc.) and animal (including fishes) components together. The IFS combines crop, livestock, aquaculture, and agro-industry as a symbiotic system. Recycling of resources is the basic principle of IFS. That means the wastes of one process/component become the input for others. There are various concepts, strategies, and practices in IFS that are followed throughout the world but the common characteristic of those systems is that they have a combination of crop and livestock components including trees and aquaculture [4, 34]. The IFS built up with the concept of combing site-specific crop + animal components, for minimizing system risk, enhancing system-productivity, reducing external input use, recycling of bio-resources, and crop residues. Broadly, the goals of IFS are to ensure stable income, maintain agro-ecological equilibrium, reduce fertilizers and pesticide use, effective utilization of organic resources within the farm boundary, and conserve

Table 7.1 Crop management practices contribute to climate change (CC) adaptation or mitigation or both

Crop management practices contribute to CC adaptation	Crop management practices contribute to CC mitigation	Crop management practices contribute to both CC adaptation and mitigation
Conservation agriculture (CA)	Conservation agriculture (CA)	Conservation agriculture (CA)
Site-specific agro-ecosystem-based approaches	Site-specific agro-ecosystem based approaches	Site-specific agro-ecosystem based approaches
Integrated soil and nutrient management	Lowering the N-fertilizer use (urea-briquette deep placement)	Integrated nutrient management (INM)
	Integrated nutrient management (INM)	
	Slow-release N-fertilizer and use of nitrification inhibitors	
Mulching and cover cropping	Mulching and cover cropping	Mulching and cover cropping
Change in cropping sequence/rotations	Introduction of legumes in cropping system	–
Crop-diversification	Crop-diversification	Site-specific farming systems with diversified cropping system
Use of high quality and stress (climatic) tolerant seeds of adapted varieties	Cultivating varieties with higher NUE and WUE	Farming with cultivars having stress tolerance and better physiological NUE/WUE
Integrated pest, and weed management	–	–
Management of water and irrigation systems	Management of water table	Participatory watershed management and water conservation (runoff and irrigation
Pollination management at landscape level	–	–
Site-specific organic agriculture for horticulture, spices and high-value crops	–	–
Land-fragmentation	Wet and peatland restoration	Agroforestry
–	Production of energy crops in waste/degraded lands to replace fossil fuel use	–
–	Emission reduction from animal waste/manures	–
–	Location-specific introduction of Improved SRI (system of rice intensification)	–

Table 7.2 Major abiotic stress-tolerant varieties in South Asia

Abiotic Stress	Varieties	Country
Rice		
Flood and submergence	Swarna Sub-1; IR-64 Sub1; CR Dhan-505; (iv) CR Dhan-500; Jalamani; VarshaDhan	India
Salinity	Nuna Subarna	India
Drought	Sahabhagi Dhan; Kalinga-2; Narrendra Dhan-97	India
Wheat		
Drought	HI 1531; HI 8627; HD 2888; HPW 349; PBW 527; PBW 644; WH 1080; HD 3043; PBW 396; HD 2987; NIAW 1415; K 8962; HD 4627	India
	Dharabi, FSD 08; Khirman; Ihsan	Pakistan
Heat	DBW 107; K1114; NIAW 1994	India
	Gold; AAS; Jauhar, Galaxy; Ujala	Pakistan
Salinity	KRL 1–4; KRL 210; KRL 213; KRL 19	India
	Pasban; Sehr; Uqab	Pakistan
Maize		
Water logging, submergence, deep-water	T 5084; Seed Tech-1314; HM 5; HM 10; PMH 2	India
Drought	HM 4; Pus Hybrid Makka 5; Pusa Hybrid Makka 1; DHM 121	India
Heat	Lall 454; RCRMH 2	India
	BHM 14; BHM 15	Bangladesh
	YH 1898; FH 793; KJ Surabhi; ND 6339; NK 64,017	Pakistan
	Rampur Hybrid 10; Rampur Hybrid 8	Nepal
Cold and frost	HHM 1; HM 1; HQPM 1	India

Source Aryal et al. [2]

natural resources for protecting the environment in long run. If we see critically, the goals of IFS are perfectly matching with the three objectives/pillars of CSA (productivity enhancement, system resilience, and climate change mitigation) (Fig. 7.3). The IFS is inherently "climate smart." However, the ecosystem services provided, the resilience imparted into the system, the extent of climate-proofing done, and the carbon sequestered by IFS are poorly quantified and documented.

The IFS is a "climate proof" technology and imparted intrinsic climate-resilience (the 2nd pillar of CSA) to the systems. The benefits obtained from IFS which increase the resilience of the system can be listed as, (i) provides diversified products and reduces risk of market fluctuations of one/single product; (ii) improves

Table 7.3 Agroecology-based rice management practices for CSA

Ecology	Stress tolerant cultivars (e.g., from India)	Production and protection technology	People participation/community support
Rainfed-upland ecology	Kalinga-2; Sahabhagi; Dhan-97; Narendra	(i) In-situ moisture conservation (ii) DSR (iii) Site-specific INM (iv) IPM	(i) Rainwater harvesting' (ii) Water harvesting and recycling (iii) Field/contour bunding (iv) Supplemental irrigation
Rainfed-lowland ecology	(i) Swarna (ii) CRDhan-500 (iii) Pooja (iv) Ranjit (v) Maudomani	(i) DSR; INM; IPM (ii)Weather forecasting based real-time pest management	(i) Community seed bank (ii) Community nursery (iii) Custom hiring of high-end farm equipment
Irrigated-ecology	(i) Hybrids (ii) Site-specific HYV	(i) DSR; INM; IPM (ii) Field-drainage (iii)Weather forecasting based real-time pest management	(i) Community seed bank (ii) Community nursery (iii) Custom hiring of high-end farm equipment
Submergence and flood-prone ecology	(i) Sub-1-introgressed cultivars (e.g. Swarna Sub-1; IR64 Sub1) (ii) CR Dhan-505 (iii) Jalamani (iv) Varsha Dhan	(i) Integrated Drainage networking (ii) Deep placement of urea briquette (iii) Foliar application of nutrients and pesticides	(i) Community seed bank (ii) Community nursery
Coastal-saline ecology	'*Luna Subarna*' '*Luna Shanki*'	(i) Surface and sub-surface drainage (ii) INM; IPM (iii) Irrigation with good quality water	(i) Community nursery (ii) Field bunding (iii) Cyclone-risk avoidance mechanism

Table 7.4 Annual greenhouse gas emission from lowland flooded rice after 39 years of INM practices

Treatment	CH_4 emission (kg ha^{-1})	CO_2–C emission (kg ha^{-1})	N_2O–N emission (kg ha^{-1})
NPK (inorganic only)	124.1^a	1210.1^a	1.46^a
FYM + NPK	177.6^b	1407.0^b	1.82^b

Source Bhattacharyya et al. [7, 9]

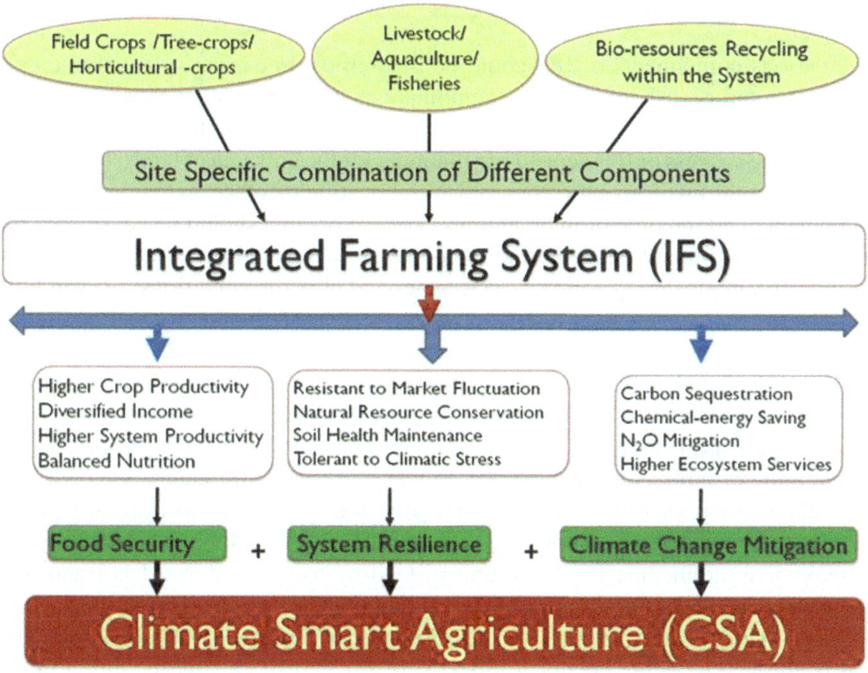

Fig. 7.3 Integrated farming system (IFS) is a climate-smart approach

soil chemical (organic carbon, fertility), physical (water holding capacity, aggregation, porosity, infiltration rate), and biological (microbial population and functions, enzymatic activities) processes; (iii) encourages less dependence of external inputs (pesticides, fertilizers, energy, feeds); (iv) assures labor supply even in crisis as it mostly uses family labor for the basic component of maintenance of the system; and (v) ensures a higher net return from diversified income sources by providing a buffer against price-fluctuations, trade-inconsistencies, and climatic-vagaries [27, 35–37, 39]. Apart from imparted strong resilience to the ecosystem the IFS also provides sustainable food and livelihood security (the first pillar of CSA) to small and marginal farmers in resource-poor countries. Therefore, the CSA should include IFS with an efficient crop, animal, water, soil, nutrient, and pest management practices, that are cost-effective and environmental friendly [64, 67, 71]. The enhanced profit margin with balanced nutrition to the farm family through on-farm production, further ensures less external-input dependence and self-sustainability of the system. Those are the key factors of the sustainability of food and nutritional security. In the ideal IFS system, the flow-dynamics of all the four primary ecosystem processes like energy, water, mineral, and bio-resources are maintained in harmony.

Components and global extent of IFS

The various components of IFS could be field-crops, tree-crops, plantation-crops, livestock, poultry, fish, etc., but the combinations of crop must be chosen carefully considering social, economic, and environmental conditions. The most important aspect of the success of IFS depends on planning for the unit area of land (should take, "farm as a unit") with specific components, allocation of space and budget for each component, and the time of interventions. The factors that should be considered for planning and compatible integration of components/farm enterprises for a location-specific IFS model are as follows: (i) weather and climatic conditions; (ii) soil type and fertility status; (iii) land topography and ecology; (iv) quantity and quality of water availability; (v) labor resource availability, both skilled and unskilled; (vi) social and cultural preference of components of IFS; (vi) type (small, marginal, or large) and vulnerability of farmers to climate change; (vii) economics of the proposed IFS; and (viii) perception and attitude of farmers toward IFS. The perception and attitude of farmers toward IFS are critically important for adaptation of IFS in developing countries of the world, particularly in tropical and sub-tropical regions. In general, poor livelihood options, having small farm holding; but farmers want to maximize the income from that unit of land by using their own labor and bio-resources, and where the water and external inputs are relatively costlier, the farmers want to venture for IFS. Further, the positive attitude of farmers toward maintaining soil health and reducing chemical use, cost of waste disposal, and pollution in the field prefer their choice toward IFS.

Integrated farming systems (IFS) have traditionally been practiced by small and marginal farmers in countries like India, China, Bangladesh, Indonesia, Vietnam, Malaysia, Rwanda, and Thailand in Asia. However, due to burgeoning demand for specific type food from mass population, these traditional systems in many places have been replaced by staple crop and cash crop systems which are also many times promoted by Government to meet the increasing demand of food [57, 59]. The few examples of IFS systems are "crop-livestock farming system"; "crop-aquaculture farming system"; "crop-poultry farming system"; "crop-fish-poultry farming system"; "crop-livestock-poultry farming system"; "crop-livestock-fish-poultry farming system"; etc. In the crop-livestock system, on-farm manure prepared from livestock could improve soil fertility, and thereby helps in balanced nutrient supply to crop from soil and maintains crop productivity. On the other hand, crop residues from different crop-component would supply feed to livestock regularly [27]. The "crop-aquaculture farming system" is prevalent in China, India, Japan, Indonesia, the Philippines, and Thailand. Fish culture improves soil fertility and inhibits weed infestation in rice crop [14, 29, 55]. However, rice also provides periphytic, planktonic, and benthic food to fish [48]. Rice fish co-culture has the potential to mitigate GHG emissions, particularly nitrous oxide (at the tune of 9%) over single rice cultivation although it promotes methane emission a bit [7, 9]. The rice-fish farming system has a huge potential in lowland rice ecologies with multifaceted benefits including climate change mitigation and needs to be prompted extensively under CSA.

The components and strategy of IFS are highly location-specific. It has different models and modes of actions with various pay-back periods. Its components are sensitive to climate and weather variables, investment potential, the skill of farmers, and cultural back-stepping of the particular locality. As a whole, it has huge potential to sustain long-term productivity and mitigate climate change with greater adaptive capacity. There is plenty of evidence that the majority of IFS models generate employment, recycle bio-resources, and enhance income opportunity to homestead but how much it combats climate change consequences and mitigate is not widely proved and documented.

7.4.4 Participatory Watershed Management and Soil-Water Conservation

Participatory watershed management represents an ideal CSA practice (Fig. 7.4). The approach is highly successful in degraded hilly regions of Asia and Africa. It is a natural resource base rehabilitation of degraded land/ecologies with the active participation of local people. The three primary interventions of this approach are rainwater

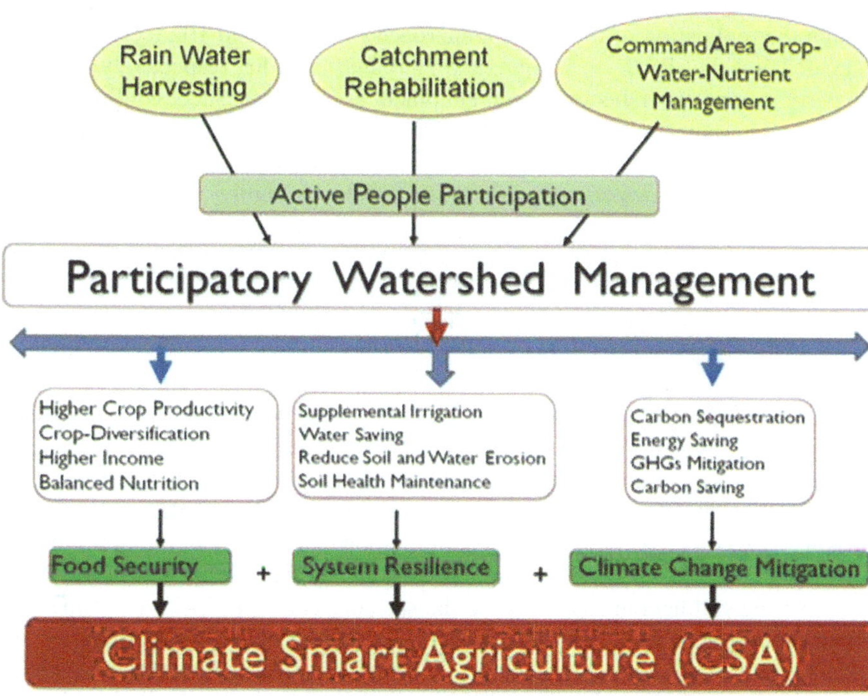

Fig. 7.4 Participatory watershed management (PWSM) is a climate-smart approach

harvesting and water resource development, catchment treatment/rehabilitation and enhancement of the agricultural (both crop and animal component) productivity. It is a typical "bottom-up" approach by involving the people at the initial planning stage itself. Being integrated in nature, the participatory watershed management ensures effective collaboration of all stakeholders, like, farmers, local institutions, and social groups. And it is being recognized as the promising approach for CSA in natural resource management, particularly in climate-vulnerable *rainfed* ecosystems. The components of watershed management include rainwater harvesting in small water harvesting structures (WHS)/ponds, or check dams; contour/graded bunding; diversion drains or runoff-drainage-network; contour/stagger trenching and afforestation in the catchment; spurs in torrents; introduction of climate-stress-tolerant high-yield varieties on command area; real-time nutrient management of crops; and maintenance of infrastructure through watershed management committee (WMC). The success story of "participatory watershed management" can be cited through the achievement of "Sukhomajri" (Haryana, India). Thirty years after the implementation (from 1977 to 2000) of the project, the B: C ratio was increased from 1.1 to 1.9. Apart from that, greenery was returned to the barren catchment; the sediment load in torrent reduced from 141 to 19 t/ha/year in degraded Shivaiks, Himalayan region, India; 3–4 time increased in crop production (maize and what) was observed, and gave high resilience to crop-livestock systems over two decades against the negative of climate. The success of the "Sukhomajri" project was established further through the replication of the concept/approach in 30, 72, and 102 locations in the states of Himachal Pradesh, Punjab, and Haryana, in India, respectively [44, 47]. The participatory watershed management practices have been successfully operated as climate-smart technology in the hilly terrain of the Himalayas in India and Nepal. It is an effective CSA and if integrates with weather forecasting, crop insurance, and incentives for saving water, soil and carbon would be highly adaptive practice in CSA.

7.4.5 Integrated Nutrient Management

The integrated nutrient management (INM) refers to "an approach of the site-specific application of organic manure, inorganic fertilizer, and biological amendments in combination to increase nutrient use efficiency and to sustain crop productivity" [6]. Simultaneously, it reduces losses of nutrients by adjusting the crop demand with soil-nutrient availability [31, 74]. The three key components of INM are, (i) optimum use of available sources of nutrients (inorganic, organic, biofertilizer, crop-residues, etc.) for sustaining soil health; (ii) matching the soil nutrient supply with real-time crop demand for maximizing crop yield; and (iii) increasing nutrient use efficiency and reducing the pollution through checking the losses. So, on principle, INM covered all the three pillars of CSA, namely, enhanced productivity, resilience, and mitigation to climate change (Fig. 7.5). The INM is an age-old practice followed, modified, and validated globally for decades, therefore, nothing new about this technology to

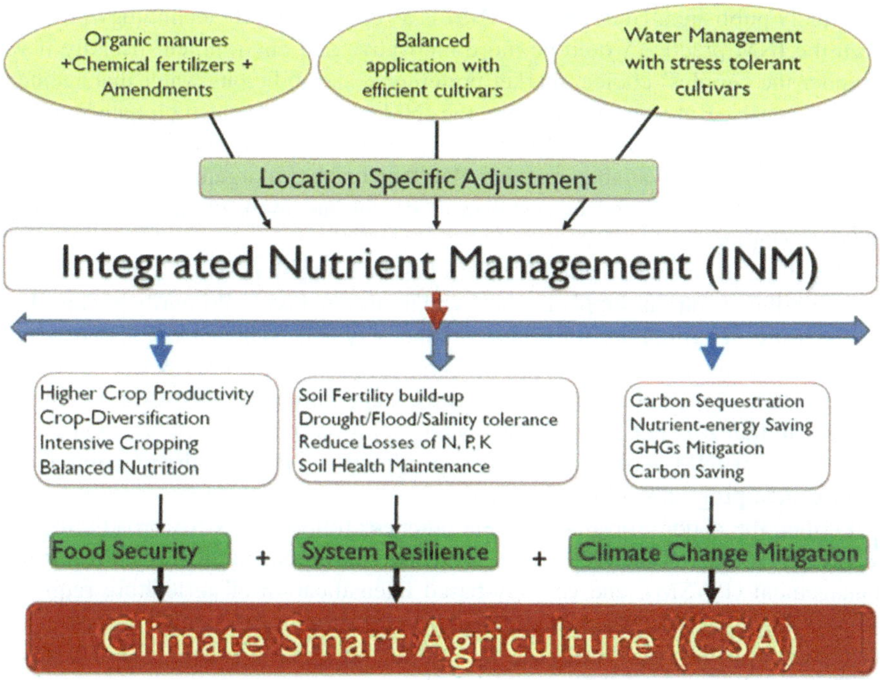

Fig. 7.5 Integrated nutrient management (INM) is a climate-smart approach

prove its efficiency. It is a part of "sustainable goal for agriculture" and recommended in many international and national policies. However, it is still equally relevant in the context of present day food security and climate-change challenges [6, 70]. In CSA, the site-specific INM strategies like a combination of inorganic and organics sources of nutrients in agriculture could be promoted into "high-input" systems for reducing chemical use and sustaining soil health and to "low-input" systems for enhancing nutrient supply and productivity. The main benefit is that the INM is flexible, widely adapted, and suitable to various kinds of cropping systems. It provides system tolerance to climate uncertainties by assuring consistent nutrient supply to crops (field-crop, horticultural-crop, grassland, pasture, tree-crops, etc.). At the same time, the consistent addition of organic manure to the system slowly releases the nutrients from fertilizer during crop growth which inhibits the losses of nitrogen and other nutrients and at the same time mitigatse the GHG emissions. The organic manure/crop residues having relatively higher C: N ratios decompose slowly than fertilizers; hence, slowly and steadily supply the nutrients to plant based on the crop-demand, and thereby reduces the losses [73]. The long-term INM practice could sequester higher carbon and reduces GHG emissions in lowland rice production systems apart from maintaining a higher yield ([7, 9], Table 7.4).

However, many studies have reported the impacts of INM on NUE, soil quality, and crop productivity, while the effect of INM on mitigation to climate change is not widely published. Undoubtedly, INM is a "climate-smart" technology/practice. Even the INM practice would be more innovative and environment-friendly if we consider the peoples' choice, market preference, and judicious water management simultaneously. But at the same time, the INM practices are also climate-sensitive and need to be modified with climate-change consequences. The use of more crop residues and locally available organic rather than costlier inorganic fertilizers should be promoted for small and marginal farmers who are more vulnerable to climate change vagaries.

Therefore, in nutshell, the INM approaches are using all possible sources of nutrients to optimize nutrient supply to plant with minimum losses. Primarily it is an adaptive measure in agriculture for higher productivity and NUE but can also generate significant mitigation effects to climate change [52]. So, it provides a "win-win" situation of higher production with climate change mitigation. However, we should be careful, that it is not a "one-size-fits-all" solution for all locations, the essence and success of INM depends on its site-specificity and demand-driven approach to development programs (Fig. 7.5) [6].

Further, the major crop-management practices followed in CSA broadly include conservation agriculture (CA), IFS, crop diversification, participatory watershed management (PWSM); and ecology-based intensification of agriculture require a suitable INM that should have significant benefits for climate change mitigation and adaptation as well.

The overall benefit of INM can be listed as,

(i) Enhance the quantity and quality of crop produce;
(ii) Reduce the losses of nutrients (major as well as minor), increase the NUE (nutrient use efficiency), and optimize the fertilizer use;
(iii) Utilize natural organics (manure, bio-fertilizers, organic-soil amendments, etc.) and enhance the use efficiency of native soil-nutrients, recycle the crop residues and maintain the soil health;
(iv) It moderates the problems like soil-salinity, soil-alkalinity, soil-acidity, and soil-compaction which hinders the crop productivity, and thereby, imparts resilience to the soil-plant system and also retards soil degradation;
(v) Mitigate GHG emission, increase carbon sequestration, and minimize energy consumption;
(vi) Sustain ecological balance and economic viability of the production system.

7.4.6 Agroforestry

Agroforestry is an intentional integration of crops, trees, and shrubs as part of agricultural systems is a paradigmatic example of "Climate-Smart Agriculture (CSA)" [10]. It facilitates water-infiltration, soil aggregation, and prevents soil erosion, and also inhibits the negative impacts of climate change. It satisfies all the three requirements of CSA, like sustaining food security, providing system resilience, and mitigating climate change (Fig. 7.6). Primarily, it enhances the system-productivity of

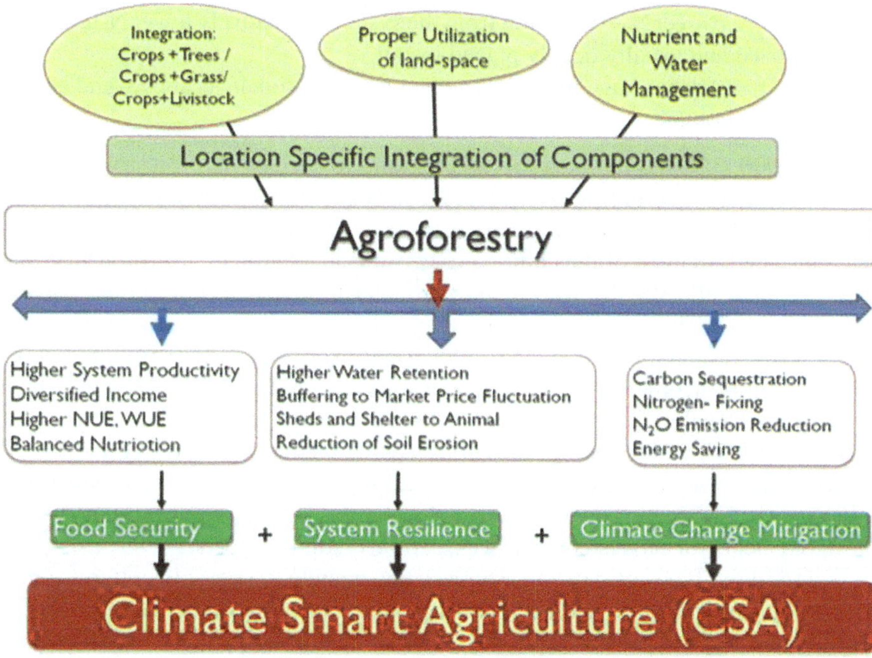

Fig. 7.6 Agroforestry is a Climate-Smart approach

crops/trees/ fodder through balanced water and nutrient supply (by N-fixing leguminous trees, e.g., *Faidherbia albida*, etc.); providing necessary energy; and improving soil fertility and crop yields [1, 25]. Secondly, it imparts the system resilience to climatic vagaries by diversifying production (tree/crops/shrubs/livestock) and providing a buffer to the system against weather-hazards. For example, tree shades protect animals from heat stress and maintain their productivity. Trees also check soil erosion and land degradation, supply quality of forage to livestock, curb overgrazing, and restore ecosystem balance [5, 6, 33, 68]. Thirdly, agroforestry also plays an important role in the reduction of GHG emissions, carbon sequestration, and climate change mitigation (Fig. 7.6).

Agroforestry delivers a wide range of ecosystem services [21, 58, 63] and socio-economic benefits [11, 25] which are listed below:

(i) Crop/tree yields are less fluctuating due to higher water retention in soil.
(ii) Soils are protected from water and wind erosions
(iii) More stabilized income through diversified farm products.
(iv) The agroforestry system is a viable substitute for slash-and-burn systems in hills-slopes.
(v) It prevents forest degradation/deforestation by reducing pressure on forests, and thereby contributes to the mitigation of GHG emissions and climate change.

(vi) It reduces direct N_2O emission from the system as relatively lower N-fertilizers are used and usually deeply placed in soil.
(vii) Agroforestry has huge potential to act as a carbon sink in its above- and below-ground biomass.

Since 2000, FAO has initiated special programs to develop agroforestry in various "North-American" countries like Nicaragua, Guatemala, Honduras, El Salvador under the food security mission. In all the cases, the productivity of land and labor were enhanced. In most of the carbon projects the adaptation of "Agroforestry" has been promoted all over the world, and particularly in Africa (e.g., in the country Mozambique, 2003) through community involvement. Agroforestry systems are widely adapted in Asia, Africa, and Latin America mostly to the "resource-poor" (low and middle income) countries [10]. It is also promoted by United Nations in their "Sustainable Development Goals" and "Climate-Smart Agriculture" programs [10, 21, 41, 63, 69]. However, the rigorous impact assessment of agroforestry under CSA projects is necessary to promote it as a key-pathway towards improving productivity, resilience, and mitigating agents. Evaluation on the light of restoring community forest, payments for ecosystems services, carbon credit is needed [22, 45, 49]. There are mainly three types of agroforestry systems, namely, (i) agrisilvicultural (crops + trees), (ii) silvipastoral systems (forestry + animals + pastures/rangelands), (iii) agrosilvopastoral (crops + trees + animals). Another two types are also under the "Agroforestry" system, that is, "mixed garden systems" (crops + trees + animals in small plots), and "multi-use-production systems," are specifically designed for watershed recharge or erosion control or for providing limited food to livestock.

Therefore, the agroforestry systems contribute several environmental services that include on-farms recycling of organic matter, huge carbon storage in their biomass (above + below ground), groundwater recharge (on- and off-site; e.g., 'Mt. Kilimanjaro can remain the 'water tower' for long'), enriching the soil by fixing atmospheric nitrogen through leguminous-N-fixing trees (via symbiotic rhizobium association), building-up healthier soils, protecting natural resources (check the soil and water erosion), helping in pollination and pest control.

7.4.7 Climate-Smart Rice Cultivation

Rice is the staple food of more than 60% of the world population. Rice cultivation acts both as a source and sinks of GHG emissions and plays a major role in climate change feedbacks. Although management practices of rice cultivation are an integral part of other management practices discussed in this chapter like CA, IFS, INM, PWSM, and agroforestry, we would discuss this in separate section of "climate-smart rice cultivation" for its specific importance of climate change adaptation and mitigation. Rice cultivation alone contributes around 12% GHG emissions globally; on the other hand, judicious water management in rice and lowland rice has good potential to act as a carbon sink [8]. Rice has a significant adaptive capacity to climatic stresses and huge genetic diversities (having more than 35,000 thousand different germplasm of rice already discovered). Even the soil and environmental growing conditions of rice varied from mountain to lower sea level including fertile river-plain, plateau, hill-terrace, sloppy terrain, low marshy land, upland, etc. Therefore, great diversities exist in its growing conditions with respect to the variation of rainfall, timing, and extent flooding, depth, and extent of water-submergence, topography, and soil type.

Apart from its inherent tolerance to abiotic and biotic stresses, various options had tried and also have been practicing to counter-balance water stresses (excess and or deficit), drought, salinity, and heat. The different options/strategies are explained in this section. For combating flooding, the cultivars with greater plasticity and photoperiod sensitivity are required. The possession of photo-sensitivity helps the plants to escape stresses of complete submergence during flowering which causes spikelet sterility and affects grain formation [60, 61]. The double transplanting of rice is an effective practice in submergence prone areas that could be adapted in photoperiod-sensitive varieties. The anaerobic seeding tolerance is another important trait to cope up with floods. The seeding of rice-seeds under the surface of flooded soil is known as anaerobic seeding. Some rice germplasm/varieties tolerant to anaerobic seeding are Dhulia, Panikekoa, Pakhira, Panikoili, Geda Khuda, and Biradia Bankoi reported from India [13]. The negative effects of complete submergence can be overcome by using submergence tolerant cultivars having "Sub 1 genes" (e.g., rice cultivars: Swarna Sub1; IR 64 Sub1) [60] and through drainage or by manipulating soils nutrients status [13, 26, 38].

Agronomic approaches along with drought-tolerant cultivars are effective strategies to cope with the stresses of water-deficit. In this context the water-saving technologies in irrigated conditions like, "alternate wetting and drying" (AWD), "saturated soil culture," and "System of Rice Intensification" (SRI) are important. In *rainfed* condition rainwater harvesting for supplemental irrigation, field/contour bunding for soil and water conservation, contingency cropping, and crop insurance are effective. Direct seeding of rice with the introduction of drought-tolerant, short duration high yielding cultivars is the key for the management of rice under water-stress conditions (deficit). Direct seeding of rice (DSR) helps to reduce water losses by facilitating earlier crop establishment compared to transplanting, and thereby checking the losses from evaporation and deep percolation. The dry direct seeding (DDS) of rice gave higher rainfall-use efficiency (RUE), affected less from water deficit, and provide better yields [28]. Apart from that introduction of sequential and or intercropping can reduce the uncertainty and risk of crop failure in drought. Introduction of salt-tolerant varieties along with water management (draining of excess salt through leaching) is the efficient management practices to cope with the hazards of the salinity in coastal areas that resulted from sea-level rise [30, 62, 65]. A maximum daily temperature is crucial for rice as it affects the spikelet fertility that is decreasing with increasing temperature. The rice-breeding to high temperature and heat recently focused on three aspects: (i) physiological-resilience: tolerance to >40 °C temperature which is found in germplasm of arid environments (Southern Iran, Pakistan or Australia); (ii) reproductive patterns (i.e., daytime of flowering): screening of rice germplasm that flower at early morning hours; and (iii)

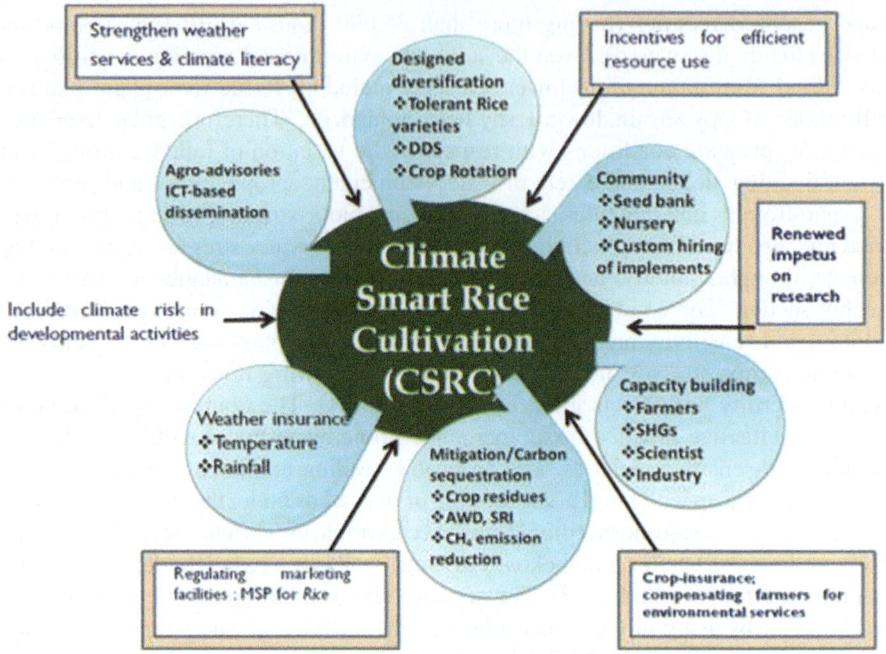

Fig. 7.7 Strategies and options of "climate-smart rice cultivation" (CSRC)

early maturity: short duration genotypes for avoiding exposure to high temperatures and terminal drought.

Now, coming to the overall technological options of "climate-smart rice cultivation" (CSRC) optimum blending of site-specific processes, a package of practices, products with weather forecasting, crop insurance, and community support (Fig. 7.7). The CSRC is something plus than "climate-resilient rice cultivation" (CRRC). So, the components include (i) introduction of climate-stress-tolerant cultivars; (ii) community seed bank and community-nursery for rice; (iii) water and nutrient management; (iv) adequate real-time supply of seed, fertilizers, equipment, protection measure through active people's participation; (v) custom hiring of climate-smart equipment like happy seeder, zero till seed drill, combine harvester, etc.; (vi) weather forecasting and advisory; and (vii) crop insurance with enabling other support systems. In this regard, it has to be noted that a decentralized government and community support system is the key to the successful adaptation of crop production practices in CSA. The aids from governmental and private institutions with innovative technologies would provide the system resilience. In this context custom hiring of farm equipment, community-based participatory seed, and fodder bank functioning at the village level are crucial. The policy-inclusive real-time crop insurance schemes based on farmers' vulnerability and advisory services must integrate with innovative technologies in CSA (Fig. 7.7).

7.5 Messages

- Crop production system provides opportunities to climate change adaptation and mitigation although it also contributes to climate change.
- The CSA demands resilient crop production practices with more efficient use of internal and external resources at the landscape level through an ecosystem approach. Therefore, only innovative technologies having a "purely technical" stand point alone would not work. They have to be integrated with active people participation and spontaneous social acceptance/support.
- Climate-smart crop management practices should also respond and adapt to changing climatic scenarios and specifically contribute to reduce GHG emissions and mitigate the future change in the climate. At the same time, the crop production system also should be productive and sustainable. So, crop production must be inherently climate smart and follow the "*save and grow*" approach.
- The genetic diversity of crops should be conserved either in-situ or ex-situ. Proper conservation and effective utilization of "crop genetic-resources" for food and agriculture not only safeguard productivity but also ensures food security and helps in coping with climate change (e.g., seeds with higher yields/better nutritional quality/earlier vigor/tolerance to abiotic and biotic stress). The "trade-transfer-transparency" (TTT) policy for sharing and utilization of crop genetic resources and knowledge among the states of the countries and nations should be encouraged and promoted.
- There is huge potential to develop climate stress-tolerant crop varieties through conventional and or molecular-breeding approaches by utilization of genetic resources.
- The CA, IFS, INM, PWSM, CSRC are well-known names of crop management practices implemented and validated throughout the world. However, site-specific innovative technologies need to be integrated with them to make that "climate smart."
- It has been well accepted now, active people participation along with community and policy supports are essential to make climate-smart crop management practices more adaptive. For example, the IFS models developed for different ecological ecosystems/sub-systems could be tuned through peoples' participatory trials with the integration of multilevel interventions at farmers' fields itself.
- Conservation of natural resources (soil, water, crop residues, and plant genetic materials) and recycling or utilization is common in all crop management practices under CSA.
- The approaches of "climate-change adaptation and mitigation" and "sustainable crop production" in the food and agriculture sector are not distinct from each other. Both try to ensure sufficient and safe nutritious food for present and future generations; a healthy and resilient ecosystem, and sustainable ecosystem services with proper conservation of biodiversity.

References

1. Amadu FO, McNamara PE, Miller DC (2020) Understanding the adoption of climate-smart agriculture: a farm-level typology with empirical evidence from southern Malawi. World Dev 126:104692
2. Aryal JP, Sapkota TB, Khurana R, Khatri-Chhetri A, Jat ML (2019) Climate change and agriculture in South Asia: adaptation options in smallholder production systems. Environ Dev Sustain 1–31
3. Beddington J, Asaduzzaman M, Fernandez A, Clark M, Guillou M, Jahn M, Erda L, Mamo T, Van Bo N, Nobre CA, Scholes R, Sharma R, Wakhungu J (2011) Achieving food security in the face of climate change: summary for policy makers from the commission on sustainable agriculture and climate change. CRP-CCAFS, Copenhagen. www.ccafs.cgiar.org/commission
4. Bhatt BP, Bujarbaruah KM, Sharma YP (2006) Integrated farming system: a sustainable alternative for the benefit of small-scale farmers and the environment
5. Bhattacharyya P, Roy KS, Nayak AK (2016) Greenhouse gas emission from agriculture monitoring, quantification and mitigation. Narendra Publishing House, Delhi India. ISBN: 978-93-84337-00-0, p 193
6. Bhattacharyya P, Munda S, Dash PK (2019) Climate change and greenhouse gas emission. New India Publishing Agency, New Delhi, 110088, India. ISBN: 978-93-895-7175-2, p 195
7. Bhattacharyya P, Nayak AK, Mohanty S, Tripathi R, Shahid M, Kumar A, Dash PK et al (2013) Greenhouse gas emission in relation to labile soil C, N pools and functional microbial diversity as influenced by 39 years long-term fertilizer management in tropical rice. Soil Tillage Res 129:93–105
8. Bhattacharyya P, Neogi S, Roy KS, Dash PK, Nayak AK, Mohapatra T (2014) Tropical low land rice ecosystem is a net carbon sink. Agr Ecosyst Environ 189:127–135
9. Bhattacharyya P, Sinhababu DP, Roy KS, Dash PK, Sahu PK, Dandapat R, Mohanty S et al (2013) Effect of fish species on methane and nitrous oxide emission in relation to soil C, N pools and enzymatic activities in rainfed shallow lowland rice-fish farming system. Agr Ecosyst Environ 176:53–62
10. Blaser WJ, Oppong J, Hart SP, Landolt J, Yeboah E, Six J (2018) Climate-smart sustainable agriculture in low-to-intermediate shade agroforests. Nat Sustain 1(5):234–239
11. Brown SE, Miller DC, Ordonez PJ, Baylis K (2018) Evidence for the impacts of agroforestry on agricultural productivity, ecosystem services, and human well-being in high-income countries: a systematic map protocol. Environ Evid 7(1):24
12. Burlingame B, Mouillé B, Charrondiere R (2009) Nutrients, bioactive non-nutrients and anti-nutrients in potatoes. J Food Compos Anal 22(6):494–502
13. Das KK, Panda D, Sarkar RK, Reddy JN, Ismail AM (2009) Submergence tolerance in relation to variable floodwater conditions in rice. Environ Exp Bot 66(3):425–434
14. Dugan P, Dey MM, Sugunan VV (2006) Fisheries and water productivity in tropical river basins: enhancing food security and livelihoods by managing water for fish. Agric Water Manag 80(1–3):262–275
15. FAO (2009) How to feed the world in 2050. Executive summary. Food and Agriculture Organization of the United Nations, Rome
16. FAO (2010) The state of food insecurity in the world 2010. Addressing food insecurity in protracted crises. Food and Agriculture Organization of the United Nations, Rome
17. FAO (2012) The state of food insecurity in the world 2012. FAO, Rome
18. FAO (2013) FAO cereal supply and demand brief. https://www.fao.org/worldfoodsituation/wfs-home/csdb/en/
19. FAO (2013) FAOSTAT emissions agriculture database. https://faostat.fao.org/site/705/default.aspx
20. FAO, IFAD, IMF, OECD, UNCTAD, WFP (2011) The World Bank, The WTO, IFPRI and The UN HLTF. 2011. Price volatility in food and agricultural markets: policy responses. Rome, FAO

21. Félix GF, Scholberg JM, Clermont-Dauphin C, Cournac L, Tittonell P (2018) Enhancing agroecosystem productivity with woody perennials in semi-arid West Africa. A meta-analysis
22. Ferraro PJ, Hanauer MM, Miteva DA, Nelson JL, Pattanayak SK, Nolte C, Sims KR (2015) Estimating the impacts of conservation on ecosystem services and poverty by integrating modeling and evaluation. Proc Natl Acad Sci 112(24):7420–7425
23. Galluzzi G, Duijvendijk CV, Collette L, Azzu N, Hodgkin T (2011) Biodiversity for food and agriculture: contributing to food security and sustainability in a changing world. In: Outcomes of an expert workshop held by FAO and the platform on agrobiodiversity research from 14–16 April 2010, Rome, Italy
24. Garnett T, Appleby MC, Balmford A, Bateman IJ, Benton TG, Bloomer P, Herrero M et al (2013) Sustainable intensification in agriculture: premises and policies. Science 341(6141):33–34
25. Garrity DP, Akinnifesi FK, Ajayi OC, Weldesemayat SG, Mowo JG, Kalinganire A, Bayala J et al (2010) Evergreen Agriculture: a robust approach to sustainable food security in Africa. Food Secur 2(3):197–214
26. Gautam P, Lal B, Raja R, Baig MJ, Haldar D, Rath L, Nayak AK et al (2014) Post–flood nitrogen and basal phosphorus management affects survival, metabolic changes and anti-oxidant enzyme activities of submerged rice (Oryza sativa). Funct Plant Biol 41(12):1284–1294
27. Gupta V, Rai PK, Risam KS (2012) Integrated crop-livestock farming systems: a strategy for resource conservation and environmental sustainability. Indian Res J Ext Educ Spec Issue 2:49–54
28. Haefele SM, Bouman BAM (2009) Drought-prone rainfed lowland rice in Asia: limitations and management options. In: Drought frontiers in rice: crop improvement for increased rainfed production, pp 211–232
29. Huy Giap D, Yi Y, Kwei Lin C (2005) Effects of different fertilization and feeding regimes on the production of integrated farming of rice and prawn Macrobrachium rosenbergii (De Man). Aquac Res 36(3):292–299
30. Ismail AM, Heuer S, Thomson MJ, Wissuwa M (2007) Genetic and genomic approaches to develop rice germplasm for problem soils. Plant Mol Biol 65(4):547–570
31. Janssen BH (1998) Efficient use of nutrients: an art of balancing. Field Crops Res 56(1–2):197–201
32. Jarvis A, Upadhyaya HD, Gowda CLL, Agrawal PK, Fujisaka S, Anderson B (2008) Climate change and its effect on conservation and use of plant genetic resources for food and agriculture and associated biodiversity for food security
33. Jonker A, Farrell L, Scobie D, Dynes R, Edwards G, Hague H, Waghorn G et al (2019) Methane and carbon dioxide emissions from lactating dairy cows grazing mature ryegrass/white clover or a diverse pasture comprising ryegrass, legumes and herbs. Anim Prod Sci 59(6):1063–1069
34. Kumar R, Patra MK, Thirugnanavel A, Deka BC, Chatterjee D, Borah TR, Upadhyay PK et al (2018) Comparative evaluation of different integrated farming system models for small and marginal farmers under the Eastern Himalayas. Indian J Agric Sci 88(11):1722–1729
35. Kumar V, Rajput P, Goel A (2018) Atmospheric abundance of HULIS during wintertime in Indo-Gangetic Plain: impact of biomass burning emissions. J Atmos Chem 75(4):385–398
36. Kumar V, Saharawat YS, Gathala MK, Jat AS, Singh SK, Chaudhary N, Jat ML (2013) Effect of different tillage and seeding methods on energy use efficiency and productivity of wheat in the Indo-Gangetic Plains. Field Crops Res 142:1–8
37. Kumar V, Sarkar C, Sinha V (2016) Influence of post-harvest crop residue fires on surface ozone mixing ratios in the NW IGP analyzed using 2 years of continuous in situ trace gas measurements. J Geophys Res Atmos 121(7):3619–3633
38. Lal B, Gautam P, Rath L, Haldar D, Panda BB, Raja R, Nayak AK et al (2015) Effect of nutrient application on growth, metabolic and enzymatic activities of rice seedlings during flooding stress and subsequent re-aeration. J Agron Crop Sci 201(2):138–151
39. Lal R, Miller FP (1990) Sustainable farming for tropics. Sustain Agric Issues Prospect 1:69–89
40. Lin BB (2011) Resilience in agriculture through crop diversification: adaptive management for environmental change. Bioscience 61(3):183–193

41. Lipper L, Thornton P, Campbell BM, Baedeker T, Braimoh A, Bwalya M, Hottle R et al (2014) Climate-smart agriculture for food security. Nat Clim Change 4(12):1068–1072
42. Loboguerrero AM, Campbell BM, Cooper PJ, Hansen JW, Rosenstock T, Wollenberg E (2019) Food and earth systems: priorities for climate change adaptation and mitigation for agriculture and food systems. Sustainability 11(5):1372
43. McDonald T, Sokolow J, Hunter D, Otieno G, Zhang W, Gowdy J, Swensson LFJ et al (2018) Farmer and community-led approaches to climate change adaptation of agriculture using agricultural biodiversity and genetic resources. Food Secur Clim Change 471–498
44. Mittal SP, Aggarwal RK, Samra JS (eds) (2000) Fifty years of research on sustainable resource management in Shivaliks. Central Soil and water conservation research and training institute, Research Centre, Chandigarh-160019, India, pp 506
45. Mohebalian PM, Aguilar FX (2018) Design of tropical forest conservation contracts considering risk of deforestation. Land Use Policy 70:451–462
46. Munesue Y, Masui T, Fushima T (2015) The effects of reducing food losses and food waste on global food insecurity, natural resources, and greenhouse gas emissions. Environ Econ Policy Stud 17(1):43–77
47. Muruganandam M, Samra JS (2000) Fish farming potentials in small water reservoirs in Shiwaliks. In: Mittal, Agarwal SPRK, Samra JS (eds) Fifty years of research for sustainable resource management of Shiwaliks. CSWCRTI, RC, Chandigarh
48. Mustow SE (2002) The effects of shading on phytoplankton photosynthesis in rice–fish fields in Bangladesh. Agr Ecosyst Environ 90(1):89–96
49. Oldekop JA, Sims KR, Karna BK, Whittingham MJ, Agrawal A (2019) Reductions in deforestation and poverty from decentralized forest management in Nepal. Nat Sustain 2(5):421–428
50. Ortiz R, Redden R, Yadav SS, Maxted N, Dulloo ME, Guarino L, Smith P (2015) The importance of crop wild relatives, diversity, and genetic potential for adaptation to abiotic stress-prone environments. Crop Wild Relat Clim Chang 80–87
51. Pandey D, Agrawal M, Bohra JS (2012) Greenhouse gas emissions from rice crop with different tillage permutations in rice–wheat system. Agr Ecosyst Environ 159:133–144
52. Pathak H, Jain N, Bhatia A, Kumar A, Chatterjee D (2016) Improved nitrogen management: a key to climate change adaptation and mitigation. Indian J Fertil 12(11):151–162
53. Pilling D, Hoffmann I (2011) Climate change and animal genetic resources for food and agriculture: state of knowledge, risks and opportunities. Background Study Paper (53)
54. Pittelkow CM, Liang X, Linquist BA, Van Groenigen KJ, Lee J, Lundy ME, Van Kessel C et al (2015) Productivity limits and potentials of the principles of conservation agriculture. Nature 517(7534):365–368
55. Poonam A, Saha S, Nayak PK, Sinhababu DP, Sahu PK, Satapathy BS, Giri SC et al (2019) Rice-fish integrated farming systems for eastern India. NRRI Research Bulletin No. 17, ICAR-National Rice Research Institute, Cuttack 753006, Odisha, India. pp 33+iii
56. Pradhan et al (2018) Potential of conservation agriculture CA for climate change adaptation and food security under rainfed upland of India: a Trans-disciplinary approach. Agric Syst 163:27–35
57. Rambo AT (2017) The agrarian transformation in Northeastern Thailand: a review of recent research. Southeast Asian Stud 6(2):211–245
58. Reed J, van Vianen J, Foli S, Clendenning J, Yang K, MacDonald M, Sunderland T et al (2017) Trees for life: the ecosystem service contribution of trees to food production and livelihoods in the tropics. For Policy Econ 84:62–71
59. Ruaysoongnern S, Suphanchaimart N (2001) Land use patterns and agricultural production systems with emphasis on changes driven by economic forces and market integration. In: Natural resource management issues in the Nakhon Ratchasima basin of NE Thailand: an overview, pp 67–77
60. Sarkar RK, Panda D, Reddy JN, Patnaik SSC, Mackill DJ, Ismail AM (2009) Performance of submergence tolerant rice (Oryza sativa) genotypes carrying the Sub1 quantitative trait locus under stressed and non-stressed natural field conditions

61. Sarkar RK, Reddy JN, Sharma SG, Ismail AM (2006) Physiological basis of submergence tolerance in rice and implications for crop improvement. Curr Sci 899–906
62. Sarker UK, Uddin MR, Sarkar MAR, Salam MA, Hasan AK (2017) Physiological parameters and yield differ in rice (Oryza sativa L.) cultivars with variable water management systems. Arch Agric Environ Sci 2(4):247–256
63. Sileshi GW (2016) The magnitude and spatial extent of influence of Faidherbia albida trees on soil properties and primary productivity in drylands. J Arid Environ 132:1–14
64. Singh SP, Gangwar B, Singh MP (2009) Economics of farming systems in Uttar Pradesh. Agric Econ Res Rev 22(1):129–138
65. Singh YP, Mishra VK, Singh S, Sharma DK, Singh D, Singh US, Ismail AM et al (2016) Productivity of sodic soils can be enhanced through the use of salt tolerant rice varieties and proper agronomic practices. Field Crops Res 190:82–90
66. Snook LK, Dulloo ME, Jarvis A, Scheldeman X, Kneller M (2011) Crop germplasm diversity: the role of gene bank collections in facilitating adaptation to climate change. Crop Adapt Clim Change 495–506
67. Sullivan CA (2011) Quantifying water vulnerability: a multi-dimensional approach. Stoch Env Res Risk Assess 25(4):627–640
68. Thornton PK, Herrero M (2010) Potential for reduced methane and carbon dioxide emissions from livestock and pasture management in the tropics. Proc Natl Acad Sci 107(46):19667–19672
69. van Noordwijk M, Duguma LA, Dewi S, Leimona B, Catacutan DC, Lusiana B, Minang PA et al (2018) SDG synergy between agriculture and forestry in the food, energy, water and income nexus: reinventing agroforestry? Curr Opin Environ Sustain 34:33–42
70. Vanlauwe B, Pypers P, Birachi EA, Nyagaya M, Schagen BV, Huising J, van Asten PJ et al (2012) Integrated soil fertility management in Central Africa: experiences of the consortium for improving agriculture based livelihoods in Central Africa (CIALCA). Centro Internacional de Agricultura Tropical (CIAT)
71. Walia SS, Kaur N (2013) Integrated farming system-an ecofriendly approach for sustainable agricultural environment-a review. Greener J Agron For Hortic 1(1):001–011
72. Walsh BS, Parratt SR, Hoffmann AA, Atkinson D, Snook RR, Bretman A, Price TA (2019) The impact of climate change on fertility. Trends Ecol Evol 34(3):249–259
73. Wortman SE, Francis CA, Bernards ML, Drijber RA, Lindquist JL (2012) Optimizing cover crop benefits with diverse mixtures and an alternative termination method. Agron J 104(5):1425–1435
74. Wu W, Ma B (2015) Integrated nutrient management (INM) for sustaining crop productivity and reducing environmental impact: a review. Sci Total Environ 512:415–427

Chapter 8
Livestock and Aquaculture Management for Climate-Smart Agriculture

8.1 Climate Change and Livestock Sector

The human population is projected to increase by 33% in 2050 with a simultaneous increase in demand for agricultural products by 60–70% including 50–60% in the livestock sector [15, 28]. The changing climatic scenario would pose a threat to the livestock sector due to adverse impact on fodder production, water scarcity, prevalence of disease and pest, and animal reproduction. This sector also supplies 33% of global protein requirements, and therefore to be managed efficiently to ensure food security [42]. At the same time, the livestock sector generates 14.5% of greenhouse gases, globally [21]. Further, in the near future, the livestock sector would be faced with challenges of land and water scarcity that would further aggravate under changing climatic conditions [14, 52]. Livestock production is projected to decrease due to poor adaptive response and more spread of diseases under changing climatic scenarios [4].

8.2 Climate Change Effect on Livestock

Global climatic change in terms of increasing temperature and atmospheric carbon dioxide levels would significantly affect the production of animal feed (Fig. 8.1). The impact would vary depending on geographical location and livestock species. Specifically, the enhanced carbon dioxide concentration would affect plant growth rate and productivity, with varying effects on C3 and C4 plants [48]. Further, the nutritional quality of fodder crops could be affected due to changes in chemical composition in plant parts, such as carbohydrate and nitrogen concentration. Higher temperatures would increase lignin content; and thereby, alter cell wall components in fodder [40]. Climatic change-induced changes in fodder quality would affect the decomposability of feed and available nutrients to animals (Fig. 8.1) [50]. Several

© The Editor(s) (if applicable) and The Author(s), under exclusive license to Springer Nature Singapore Pte Ltd. 2020

P. Bhattacharyya et al., *Climate Smart Agriculture*, Green Energy and Technology,
https://doi.org/10.1007/978-981-15-9132-7_8

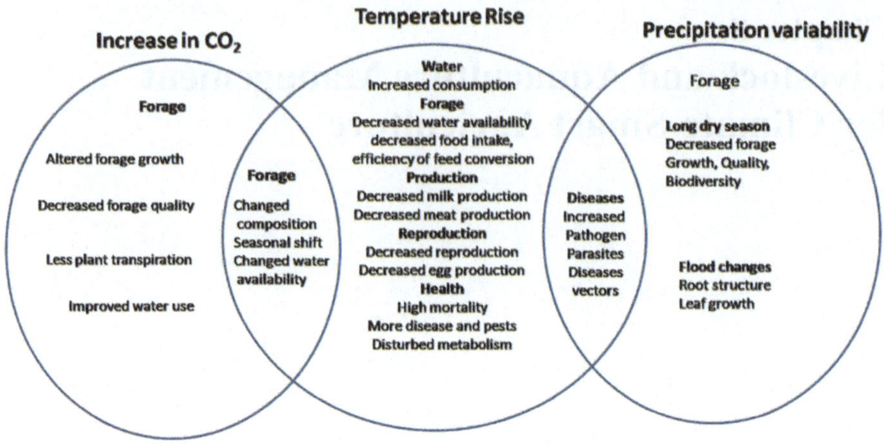

Fig. 8.1 Climate change impact on the livestock sector. *Source* Rojas-Downing et al. [41]

studies reported that climate change-induced increase in temperature could lead to a 2–3 times increase in water use by livestock itself. To face this challenge, a shift has to be made for growing fodder or rearing animals that need minimum water [36].

The impact of climate change on livestock has been found significant. A few consequences are stated in this paragraph. The increased temperature beyond the tolerable limit may increase the death rate of livestock and more so the spread of various pathogens, vectors, and parasites would be aggravated. The changing climatic parameters could cause sudden disease outbreak and might introduce new kinds of diseases, pests, pathogens in a particular area, thus, making livestock more vulnerable [50]. Moreover, the changes in rainfall patterns would significantly affect disease transmission and spread of pests like flies and mosquitoes (Fig. 8.1) [50].

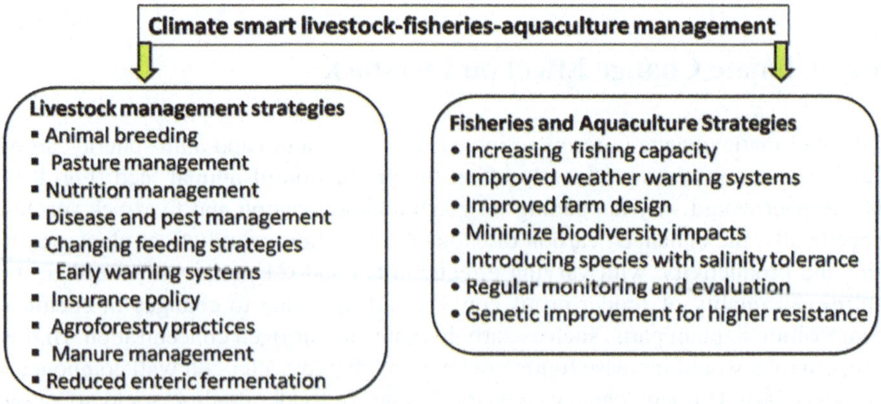

Fig. 8.2 Options for climate-smart livestock-fisheries-aquaculture management practices

The serious threat is the "heat stresses" to animals due to climate change consequences. Global climate change-induced temperature increase could affect normal physiological activities of animals and thus creating high-temperature stress in the body. This would alter the capacity of food and water consumption, rate of respiration, reproducing ability, and death rate. Many a time due to heat stress there would be the low intake of required protein, vitamins, and minerals by the animals. So they usually suffer from malnutrition that would adversely affect their metabolic functions [35]. Non-availability of sodium and potassium in food also promotes the development of a disease called metabolic alkalosis [7]. Moreover, the increased temperature decreases the reproductive efficiency of animals. Heat stress reported to reduce the sperm count concentration in bulls and adversely affects the fertility of cows [32], pigs, and poultry [30, 33].

There is a threat to change the animal biodiversity of the world due to climate change and more specifically due to an increase in global temperature. The increased temperature would affect the reproduction and mortality of a number of animal species. Under the changing climatic scenario, 15–37% of species around the globe may become extinct, causing a loss in biodiversity [46]. Therefore, in future, the animal breeds with higher climate change adaptation capacity should be identified, conserved, and promoted for sustainable livestock production.

8.3 Climate Change Effect on Milk, Fish, and Animal Production

Increase in temperature and humidity could cause up to 30% decline in milk production. In general, the increase in temperature and decrease in humidity lead to reduced feed intake and reproduction ability of animals. It affects negatively the sperm count in animals, conception rate, and ultimately triggers the calf death rate. At the same time, it decreases immunity and increases the spread of infectious diseases. In poultry, heat stress could reduce egg production and decrease the growth rate of birds. The rise in humidity favors disease outbreak.

Climate change-induced sea-level rise has considerably impacted the aquaculture and fisheries. Higher sea level brings new environmental conditions for coastal fisheries, aquaculture, and mangrove ecosystems. Extreme weather events pose threats to endanger communities residing in the coastal region, damage infrastructures, and hamper the productivity of flood-prone river belts. Moreover, the post-harvest processing, transport, and distribution would be affected, which in turn would affect the livelihood and food security.

The overall impacts of climate change on livestock, aquaculture, and fisheries are presented in tabular form below (Tables 8.1 and 8.2).

Table 8.1 Climate change impact on the livestock sector

Factor	Impacts
Water scarcity	Global climate change will worsen water scarcity scenarios. This will affect fodder production, the yield of pasture, and reduce quantity and quality of drinking water for the livestock sector
Quantity and quality of feed	i. Climate change would alter land uses for crops and fodders. This will affect the quantity and quality of feed for livestock, especially for marginal farmers ii. With variation in temperature and rainfall, the growing season will be changed, which will change the productivity and harvest index of crops iii. Increasing temperature will change the decomposition of plant material, and alter the digestibility of feed. This, in turn, reduces livestock production, the income of farmers
Biodiversity	i. An increase of 2.5 °C temperature would lead to a risk of 20–30% extinction of all plants and animals ii. Ecosystems would show a wide variation in vulnerability to changing climatic scenarios iii. Climate change would pose a threat in terms of increasing pest and disease outbreaks. This may eliminate some local and rare livestock species. Loss in biodiversity will take place due to increased health risks
Livestock health	i. More spread of vector may take place in colder regions (malaria and livestock borne diseases may shift to higher elevation) or to warmers areas (bluetongue disease in northern Europe) ii. Rainfall pattern changes may help to spread vector attack and more disease occurrence (virus of Rift Valley fever in Eastern Africa) iii. Helminth infection pattern may change due to temperature and humidity variation iv. The animal mortality rate may increase due to heat-related stress

Source Thornton et al. [49]

Table 8.2 Climate change impact on the fisheries sector

Positive impacts	i. Changing climatic scenarios may have a positive effect on the aquaculture sector by reducing cold weather impact. The growth of aquaculture species is expected to be promoted by warmer nights and winters ii. The cost for water heating and construction of ice proof structures would be lowered iii. The water not suitable for crop cultivation, such as flooded and salt-affected water may be utilized for fish production
Negative impacts	i. Changing sea level, the water level in local bodies, drought, and flood events would cause a shift in fish species across the different geographical regions. This would further affect the reproduction pattern of fish as well as nursery fish production ii. Global warming would affect pests, pathogens, and disease transmission behavior in fish and aquaculture

Source Kibuka-Musoke [31]

8.4 Climate-Smart Technologies for Livestock, Aquaculture, and Fisheries

Many adaptation and mitigation measures are to be adopted to make the livestock sector more resilient under changing climatic scenarios and make it "climate smart."

8.4.1 Livestock Management Systems and Adaptation Measures

i. Best adaptation methods could be crop and animal diversification in production systems, by incorporating agriculture and forestry production systems together and optimizing the timing of various field operations [27]. A diversified system is more tolerant to temperature and water stress and also to pest and disease attacks [3].
ii. Establishment of suitable agroforestry systems with an optimum combination of trees, crops, and pastures would improve crop production, protect the environment, and favor carbon sequestration. An integrated crop-livestock-based system would be a better measure to improve food security [24].
iii. Modification of feeding techniques is an important adaptation measure for improving livestock productivity. Few such practices are altering feed composition, the timing of feeding, quality of feed, and capacity building of produce in various locations [27]. Time to time change in a specific location for cropping and livestock production may help to prevent soil degradation and improve soil and moisture conservation [34].
iv. Another adaptation measure is to alter crop rotation; timing of field and intercultural operations to synchronize with rainfall and temperature variability [27, 34].

Breeding strategies

Breeding strategies should focus on increasing temperature tolerance, disease resistance of animals, and better reproduction traits. A suitable policy should be in place to facilitate favorable adaptation strategies. In the line of CGIAR gene banks, developing global gene banks for animals may serve as insurance under changing climatic scenarios [49].

i. The breeding method should focus on the survival of livestock under harsh conditions, such as nutrition deficiency, heat tolerance, resistance to pests, parasites, and disease [25].
ii. Identification of native livestock breed with high adaptability.
iii. Crossbreeding local livestock with high-temperature and disease-resistant breeds.

Capacity building of farmers

i. The producers should be trained for rearing livestock under water and nutrient stress conditions.
ii. Training may be provided for fodder production.

Efficient livestock management

i. Construction of shed and water sources at low cost to minimize heat stress.
ii. Keeping high-quality breeds in a lesser number will lead to resource conservation and lower greenhouse gas emission.
iii. Introduction of micro- and drip irrigation systems.
iv. Construction of water harvesting structures.

8.4.2 Livestock Management Systems and Mitigation Measures

The livestock production system contributes significantly to greenhouse gas emissions in three different ways. These are enteric fermentation, manure/waste generation, and fodder production [12, 14]. Indirectly, the livestock production system affects land-use changes and sometimes promotes deforestation due to pasture production. For example, in the "Amazon rainforest," seventy percent deforestation was attributed due to the conversion of forest to pastures [10]. Therefore, the mitigation of greenhouse gas emissions from livestock needs better feeding and manure management. Increasing the efficiency of livestock for converting energy from feed and reducing losses through waste could also help to mitigate climate change. Location-specific systematic grazing may be more effective in reducing GHG emissions and storing more soil carbon (Fig. 8.2) [45].

8.5 Climate-Smart Aquaculture Management Systems

Globally, seafood is one of the most promising international trade commodities [18]. In two hundred countries of the world, around 600 aquatic species are cultured on regular basis and the production of these aquaculture species are governed by ambient climatic conditions [19], and therefore this aquatic production sector is highly vulnerable to climate change.

A large number of aquaculture species are being affected by climate change, which necessitates the suitable adaptation options, depending on species and environmental conditions. These adaptation options are broadly classified as diet, genetic-management, relocation, information, and communication, etc., which are described below.

Diet based adaptation options

i. High-quality diet in adequate quantity would enhance the adaptation capacity under climatic stress. It has been reported that under high-temperature conditions, there were improved immune parameters of carp *Cyprinus carpio* and antioxidant enzymes, and gene expression also showed improvement [26]. The incorporation of dietary stimulants enhances the protection against adverse environmental conditions [51].
ii. There exist many aquatic species that are dependent on the environment for nutrition, like sea weeds and shellfish. Thus, climate change directly impacted the nutrient availability of those species, and the only adaptation option for those is shifting or relocation of the farm [44].

Genetics and biotechnological approaches

i. The selective breeding intervention has been reported to cause 7–12% weight gain in case of a number of carp species like *Puntius gonionotus*, *Cyprinus carpio,* and several species of prawns (*Macrobrachium rosenbergii*) [37].
ii. Retention of maximum genetic variability should be a top priority in any genetic improvement program. To date, a high success rate has been found in genetic improvement projects of *tilapia*, *prawn,* and *carps*, which is due to their wide genetic base and variability in the population [38].
iii. Resistant to disease and survival under a stressed environment needs the conservation of rare alleles associated with stress tolerance to introgression of resistance to that particular species [23]. Another approach is biological insurance through germplasm banking, which should be used in future breeding programs under CSA [2].

Management approaches

i. Adoption of best management practices against floods through engineering measures for the protection of tanks and inland infrastructures is essential to save aquaculture species under extreme events in climate change scenarios [39].
ii. The escape of species under flood situations may be prevented by placing string net structures such as those adopted for shrimps in Bangladesh [29] and Indian ponds [1].

Relocation of farm

i. The plan for emergency relocation of farm based on aquaculture zoning and risk planning should be chalked out well in advance to meet any unforeseen circumstances, to minimize the impact of climate change exposure [18].
ii. Ponds with deeper water provide higher dissolved oxygen and are more resilient under drought in dry areas [44].
iii. Aeration technique and use of oxygen tablets also favor better survival under stress conditions [1].

Diversification

Diversification of culture is itself a contingency measure under adverse conditions against losses. The practice of co-culture offers multiple benefits of enhancing nutrient use, water use, and also improves water quality [39].

Improving information, monitoring, warning, and prediction systems

i. To adopt best management practices, there is a need to retrieve timely information, forecast, and long-term prediction. Capacity building of the local community and aquaculture stakeholders is a must to improve resilience under adverse conditions [39].
ii. There is an urgent need to establish early warning technologies to minimize disaster risks such as pathogenic contamination due to extreme climate [5].
iii. Fast identification of a disease followed by quick management intervention can minimize disease impact [22].
iv. Modern approaches like web-based disease database management, tracking disease spread, surveillance, and GIS-based models are encouraged to protect aquaculture production [6] (Table 8.3).

8.6 Possibilities and Limitation of Climate-Smart Livestock Management

Given tremendously increasing demand for livestock products, to sustain the livestock production sector in the long run, it is essential to increase feed to food conversion rate. To increase the input use efficiency, the meat/milk produced per unit amount of feed or nutrients supplied is to be maximized. This will in turn increase higher crop and water productivity, higher nutrient use efficiency from manures and fertilizers, reduced emission of greenhouse gases, and decreased loss throughout the food chain [53]. The major objective is to control methane emission from enteric fermentation. The optimization of grazing practices will help to increase the productivity of grassland and pastures; however, the effect depends on climate, soil, and management practices [43]. Prevention of soil erosion and restoration of eroded lands further add to mitigation benefits [8]. The time and intensity of grazing also have to be optimized under a particular set of conditions. The practice of rotational grazing promotes higher growth and ensures better quality, and also reduces methane emission [13]. Native, as well as improved varieties of pastures with higher proportion of digestible forages should be sown. For example, in Latin America, under tropical grazing conditions, it has been shown that marked improvement in soil health, carbon stock, and productivity of farm is possible by substituting natural "*Cerrado*" vegetation with "*Brachiaria* pasture" [47]. Adoption of the best nutrients, water, and crop management practices may increase pasture quality, productivity, and performance of livestock.

The mixed system of livestock management offers multiple benefits and better adaptation to climate change by improving crop and livestock production and the reduction of greenhouse gas emissions. Integrated management of water, soil, and plant resources will increase input use efficiency and favor climate change adaptation and sustain productivity. Improved grass and fodder species are to be included in the mixed farming system. The increasing diversification of the production system will build resilience to climate change. Large-scale production of manure from animals and its anaerobic storage leads to methane emission in the atmosphere. Therefore, the storage structures must be covered to mitigate the emission of greenhouse gases. The methane inside the storage structures should be used for energy generation purposes [20]. The improving efficiency of feed conversion is an important mitigation option. Decreasing the quantity of feed needed per unit of production, such as meat or milk, will increase economic benefit and reduce the emission of greenhouse gases. Animal breed with faster growth rate and high-stress tolerance capacity having higher feed conversion efficiency should be promoted in CSA.

The transformation of rural livelihood along with livestock production system would offer better adaptation options in the livestock sector. Replacement of livestock species, integrated livestock farming systems are the need of the hour [11]. The following Table 8.4 summarizes some possible options along with limitations for climate change adaptation in the livestock sector.

Table 8.3 Strategies to reduce vulnerability in the fishery and aquaculture sector

	Parameter affected	Response
Fisheries	Decreased productivity	i. Increasing fishing capacity ii. Expand targeted aquatic species iii. Increase feeding efficiency iv. Diversification of livelihood options
	Fluctuation in yield level	i. Diversification of livelihood activities ii. Adoption of insurance schemes iii. Best management frameworks
	Sea level changes	i. Improved defense for infrastructures ii. Rehabilitation of affected community iii. Integrated management of coastal areas iv. Early weather warning mechanism
	Increased dangers of fishing	i. Weather warning mechanism ii. Improved communication and safety
	Social security	i. Local institutional arrangements

(continued)

Table 8.3 (continued)

	Parameter affected	Response
Aquaculture post-harvest	Extreme climatic events	i. Improved farm design ii. Suitable insurance policy iii. Use of indigenous breeds
	Rise in temperature	i. Efficient water management ii. Selective breeding iii. Genetic improvement iv. Best market schedules
	Water deficit	i. Shifting to coastal aquaculture ii. Culture fisheries iii. A short cycle of production iv. Better seedling quality
	Sea level rise	i. Shifting of species ii. Introduction of salt-tolerant species iii. Use of hatchery seed
	Eutrophication	i. Better planning ii. Routine monitoring iii. Emergency measures
	Increased Disease and pests	i. Improved stress management ii. Biosafety and security measures iii. Breeding for higher disease resistance
	Acidification impact	i. Alternate production method ii. Alternate handling mechanism iii. Species selection
	Effects on Infrastructure	i. Early weather warning ii. Improved defense mechanism iii. Rehabilitation iv. Disaster preparedness
	Fluctuating yield, supply-demand scenarios	i. Replacement of species ii. Reduction of losses iii. Improved communication system iv. Diversification of livelihood
	Impact on processing	i. Better weather forecasting ii. Improved information and communication technologies
	Market shocks	i. Diversification of market ii. Diversification of products

Source Daw et al. [9], De Silva and Soto [10]

Table 8.4 Adaptation options in livestock sector

Adaptation Options	Limitations to Adoption
Improvement in food quality and availability i. Regeneration of native pasture, rehabilitation of degraded pasture lands, irrigation facilities, and controlled grazing ii. Incorporation of feed and nutrient supplements, optimizing stocking rate, cultivation of perennial fodder crops iii. Efficient fertilizer management, rotational grazing, and prevention of fire hazard **Land utilization and water use** i. Altering use of farmland, irrigation (US and Australia systems) ii. Enclosure for community lands (systems in Ethiopia) **Reducing heat stress** i. Providing shade and cooling facilities ii. Improved ventilation iii. Access to water resources like rivers and pools iv. Heat tolerant livestock breeds **Livestock management** i. Adopting livestock healthcare practices ii. Managing pests, weeds, and disease vectors iii. Altering patterns of breeding and reproduction iv. Improvement in water and soil management practices	**Rangelands** i. Biophysical constraints ii. Social and economic barriers iii. Uncertainty of climate **Mixed systems** i. Information deficit ii. Modeling crop-livestock system is not adequate iii. Spatial and temporal variability iv. Biophysical, socio-economic complex v. Lack of capacity building vi. Limited efficient labor resource vii. Insecure access to capital viii. Inadequate market infrastructure
Modernization of farm operations i. Forecasting infectious diseases ii. In time warning systems iii. Developing resistant infrastructure **Policy reforms** i. Insurance schemes (flood/storm) ii. Reducing market risk iii. Trade for livestock products iv. Decision support systems v. Social learning tools	i. Lack of climate information ii. Poor diseases prevention iii. Market deficit iv. Policy issues v. Genetic assessment program doesn't focus on the low input system
Diversification i. Livelihood income options ii. Livestock herd mixing iii. Multi-species herding	i. Weak institutional arrangement ii. Poor farm management planning

Source Modified from Escarcha et al. [14]

8.7 Possibilities and Limitation of Climate-Smart Aquaculture and Fishery Management Systems

The fishery and aquaculture sector together provides income to 10–12% of the global population and contributes to livelihood security. It has been estimated that more than 85% of world fish production is directly consumed as food, which supplies 15% of global protein and nutrients for around 4.3 billion people [17]. An aquatic production system consists of a minimum of 27,000 species of fish, aquatic plants, and organisms, which have been identified so far [16]. With the tremendous increase in the world population, demand from this sector will also increase. To meet the huge demand, this sector should double production shortly. However, the level of production would be affected by changing temperature, rainfall, and limited natural resources. Further development of this sector will need sound technological interventions, economic reform, and targeted policy.

These sectors are also confronted with the challenge of climate change. In order to ensure the sustainability of this sector against climate change, a number of measures need to be undertaken. The use efficiency of natural resources per unit of fish and aquatic production has to be increased. Moreover, the vulnerability of this sector is to be reduced and the resilience of communities involved in managing this sector should be strengthened. Suitable information and communication system is to be developed at the local level to favor adaptation planning. Other practices include better feeding management, pest and disease control, minimizing post-harvest losses, and development of local market and trade policies.

The various management strategies include (i) ensuring the water availability that would prevent planned migration of cattle to forest areas or irrigated regions. Till 2000, in *Kutch*, cattle migration was practiced due to lack of water, but, since water supply from the Narmada dam started, the migration is reduced; (ii) the animal produced by crossbreeding of European-American origin breeds, are sensitive to heat stress. So, the native cattle breed with high heat tolerance may be selected for breeding purposes. For example, *Geer* and *Kankreji* cattle are suitable under climate-stress conditions; (iii) promoting the biogas plants is another option. Cattle wastes can be converted to biogas inside these plants. This could further be used as cooking gas or other energy sources; (iv) and last but not the least the landless animal farmers' need planned social protection to reduce their vulnerability.

References

1. Adhikari S, Keshav CA, Barlaya G, Rathod R, Mandal RN, Ikmail S et al (2018) Adaptation and mitigation strategies of climate change impact in freshwater aquaculture in some states of India. J Fish Sci Com 12(1):16–21
2. Barrento S, Camus C, Sousa-Pinto I, Buschmann AH (2016) Germplasm banking of the giant kelp: our biological insurance in a changing environment. Algal Res 13:134–140

3. Batima P, Bat B, Tserendash L, Bayarbaatar S, Shiirev-Adya S, Tuvaansuren G, Natsagdorj L, Chuluun T (2005) Adaptation to Climate Change, vol 90. ADMON Publishing, Ulaanbaatar
4. Bett B, Kiunga P, Gachohi J, Sindato C, Mbotha D, Robinson T et al (2017) Effects of climate change on the occurrence and distribution of livestock diseases. Prev Vet Med 137:119–129
5. Bondad-Reantaso MG, Garrido-Gamarro E, McGladdery SE (2019) Climate change-driven hazards on food safety and aquatic animal health. Impacts of climate change on fisheries and aquaculture. In: Barange M, Bahri T, Beveridge MCM, Cochrane KL, Funge-Smith S, Poulain F (eds) Impacts of climate change on fisheries and aquaculture: synthesis of current knowledge, adaptation and mitigation options. Fish Aquacult Tech Pap 627. FAO, Rome, pp 517−534
6. Brugere C, Onuigbo DM, Morgan KL (2017) People matter in animal disease surveillance: challenges and opportunities for the aquaculture sector. Aquaculture 467:158–169
7. Chase LE (2006) Climate change impacts on dairy cattle. In: Fact sheet, climate change and agriculture: promoting practical and profitable responses. https://climateandfarming.org/pdfs/FactSheets/III.3Cattle.pdf
8. Conant RT, Paustian K (2002) Potential soil carbon sequestration in overgrazed grassland ecosystems. Global Biogeochem Cycles 16(4):90–91
9. Daw T, Adger N, Brown K, Badjeck MC (2009) Climate change and capture fisheries: potential impacts, adaptation and mitigation. In: Cochrane K, Young CD, Soto D, Bahri T (eds) Climate change implications for fisheries and aquaculture: overview of current scientific knowledge. FAO Fisheries and Aquaculture Technical Paper No. 530. Rome, FAO, pp 107–150, 212 pp. www.fao.org/docrep/012/i0994e/i0994e00.htm
10. De Silva SS, Soto D (2009) Climate change and aquaculture: potential impacts, adaptation and mitigation. In: Climate change implications for fisheries and aquaculture: overview of current scientific knowledge. FAO Fisheries and Aquaculture Technical Paper, 530, pp 151–212
11. Derner J, Briske D, Reeves M, Brown-Brandl T, Meehan M, Blumenthal D, Hendrickson J (2018) Vulnerability of grazing and confined livestock in the Northern Great Plains to projected mid-and late-twenty-first century climate. Clim Change 146(1–2):19–32
12. Dourmad JY, Rigolot C, Van Der Werf H (2008) Emission of greenhouse gas, developing management and animal farming systems to assist mitigation. Livest Glob Clim Change 36
13. Eagle AJ, Olander LP, Henry LR, Haugen-Kozyra K, Millar N, Robertson GP (2012) Greenhouse gas mitigation potential of agricultural land management in the United States: a synthesis of literature. Report NI R 10-04, Third Edition. Durham, USA, Nicholas Institute for Environmental Policy Solutions, Duke University
14. Escarcha JF, Lassa JA, Zander KK (2018) Livestock under climate change: a systematic review of impacts and adaptation. Climate 6(3):54
15. FAO (Food and Agriculture Organization of the United Nations) (2009) Global agriculture towards 2050. High Level Expert Forum Issues Paper, FAO, Rome
16. FAO (2010) FAO and aquatic biodiversity. Information brief. ftp://ftp.fao.org/docrep/fao/010/i0112e/i0112e17.pdf
17. FAO (2012) SOFIA: the state of world fisheries and aquaculture. FAO, Rome. 209 pp. https://www.fao.org/docrep/016/i2727e/i2727e00.htm
18. FAO (2016) The state of the world fisheries and aquaculture 2016: contributing to food security and nutrition for all. FAO, Rome
19. FAO (2018) The state of world fisheries and aquaculture 2018: meeting the sustainable development goals. FAO, Rome
20. Gerber PJ, Carsjens GJ, Pak-uthai T, Robinson TP (2008) Decision support for spatially targeted livestock policies: Diverse examples from Uganda and Thailand. Agric Syst 96(1–3):37–51
21. Gerber PJ, Steinfeld H, Henderson B, Mottet A, Opio C, Dijkman J et al (2013) Tackling climate change through livestock: a global assessment of emissions and mitigation opportunities. Food and Agriculture Organization of the United Nations (FAO)
22. Groner ML, Maynard J, Breyta R, Carnegie RB, Dobson A, Friedman CS, Noble RT et al (2016) Managing marine disease emergencies in an era of rapid change. Philos Trans R Soc B Biol Sci 371(1689):20150364

23. Gurney-Smith HJ, Wade AJ, Abbott CL (2017) Species composition and genetic diversity of farmed mussels in British Columbia, Canada. Aquaculture 466:33–40
24. Herrero M, Thornton PK, Notenbaert AM, Wood S, Msangi S, Freeman HA et al (2010) Smart investments in sustainable food production: revisiting mixed crop-livestock systems. Science 327(5967):822–825
25. Hoffmann AA, Rieseberg LH (2008) Revisiting the impact of inversions in evolution: from population genetic markers to drivers of adaptive shifts and speciation? Annu Rev Ecol Evol Syst 39:21–42
26. Huang JF, Xu QY, Chang YM (2015) Effects of temperature and dietary protein on gene expression of Hsp70 and Wap65 and immunity of juvenile mirror carp (C yprinus carpio). Aquac Res 46(11):2776–2788
27. IFAD (International Fund for Agricultural Development) (2010) Livestock and climate change. https://www.ifad.org/lrkm/events/cops/papers/climate.pdf. In: Iglesias A, Avis K, Benzie M, Fisher P, Harley M, Hodgson N, Horrocks L, Moneo M, Webb J (2007) Adaptation to climate change in the agricultural sector. AEA Energy and Environment and Universidad de Politécnica de Madrid
28. IPCC (2019, August). Land: an IPCC special report on climate change, desertification, land degradation, sustainable land management, food security, and greenhouse gas fluxes in terrestrial ecosystems. 2019. In: The approved summary for policymakers (SPM) was presented at a press conference on, vol 8
29. Kais SM, Islam MS (2018) Impacts of and resilience to climate change at the bottom of the shrimp commodity chain in Bangladesh: a preliminary investigation. Aquaculture 493:406–415
30. Karaca AG, Parker HM, Yeatman JB, McDaniel CD (2002) Role of seminal plasma in heat stress infertility of broiler breeder males. Poult Sci 81(12):1904–1909
31. Kibuka-Musoke D (2007) Issue paper II. OECD, African Partnership Forum
32. King JM, Parsons DJ, Turnpenny JR, Nyangaga J, Bakari P, Wathes CM (2006) Modelling energy metabolism of Friesians in Kenya smallholdings shows how heat stress and energy deficit constrain milk yield and cow replacement rate. Anim Sci 82:705–716
33. Kunavongkrit A, Suriyasomboon A, Lundeheim N, Heard TW, Einarsson S (2005) Management and sperm production of boars under differing environmental conditions. Theriogenology 63(2):657–667
34. Kurukulasuriya P, Rosenthal S (2013) Climate change and agriculture: a review of impacts and adaptations. Climate Change Series Paper No. 91, World Bank, Washington DC
35. Mader TL (2003) Environmental stress in confined beef cattle. J Anim Sci 81(14 Suppl 2):E110–E119
36. Nardone A, Ronchi B, Lacetera N, Ranieri MS, Bernabucci U (2010) Effects of climate changes on animal production and sustainability of livestock systems. Livest Sci 130(1–3):57–69
37. Nguyen LA, Verreth JAJ, Leemans HBJ, Bosma RH, De Silva S (2016) A decision tree analysis to support potential climate change adaptations of striped catfish (Pangasianodon hypophthalmus Sauvage) farming in the Mekong Delta, Vietnam. Tropicultura 34(Special):105–115.
38. Nguyen NH (2016) Genetic improvement for important farmed aquaculture species with a reference to carp, tilapia and prawns in Asia: achievements, lessons and challenges. Fish Fish 17(2):483–506
39. Oyebola OO, Olatunde OM (2019) Climate change adaptation through aquaculture: ecological considerations and regulatory requirements for tropical Africa. In: Agriculture and ecosystem resilience in Sub Saharan Africa. Springer, Cham, pp 435–472
40. Polley HW, Briske DD, Morgan JA, Wolter K, Bailey DW, Brown JR (2013) Climate change and North American rangelands: trends, projections, and implications. Rangel Ecol Manag 66(5):493–511
41. Rojas-Downing MM, Nejadhashemi AP, Harrigan T, Woznicki SA (2017) Climate change and livestock: impacts, adaptation, and mitigation. Clim Risk Manag 16:145–163
42. Rosegrant MW, Fernandez M, Sinha A, Alder J, Ahammad H, Fraiture CD et al (2009) Looking into the future for agriculture and AKST. In: McIntyre BD, Herren HR, Wakhungu J, Watson RT (eds) International assessment of agricultural knowledge, science and technology

for development (IAASTD). Agriculture at a crossroads, Island Press, Washington, DC, pp 307–376

43. Smith P, Martino D, Cai Z, Gwary D, Janzen H, Kumar P et al (2008) Greenhouse gas mitigation in agriculture. Philos Trans R Soc B Biol Sci 363(1492):789–813
44. Soto D, Ross LG, Handisyde N, Bueno PB, Beveridge MC, Dabbadie L, Pongthanapanich T et al (2019) Climate change and aquaculture: vulnerability and adaptation options. *Impacts of climate change on fisheries and aquaculture*. In: Barange M, Bahri T, Beveridge MCM, Cochrane KL, Funge-Smith S, Poulain F (eds). Impacts of climate change on fisheries and aquaculture: synthesis of current knowledge, adaptation and mitigation options. Fish Aquacult Tech Pap 627. FAO, Rome, pp 465–490
45. Tennigkeit T, Wilkes A (2008) Carbon finance in rangelands. In: ICRAF/UNFCC challenges and opportunities for mitigation in the agricultural sector Technical paper
46. Thomas CD, Cameron A, Green RE, Bakkenes M, Beaumont LJ, Collingham YC et al (2004) Extinction risk from climate change. Nature 427(6970):145–148
47. Thornton PK, Herrero M (2010) Potential for reduced methane and carbon dioxide emissions from livestock and pasture management in the tropics. Proc Natl Acad Sci 107(46):19667–19672
48. Thornton PK, Boone RB, Ramírez Villegas J (2015) Climate change impacts on livestock. In: CGIAR resrach program on climate change, agriculture and food security (CCAFS), Working Paper No. 120
49. Thornton PK, Jones PG, Owiyo T, Kruska RL, Herrero M, Orindi V et al (2008) Climate change and poverty in Africa: mapping hotspots of vulnerability. Afr J Agric Res Econ 2(311–2016–5524):24–44
50. Thornton PK, van de Steeg J, Notenbaert A, Herrero M (2009) The impacts of climate change on livestock and livestock systems in developing countries: a review of what we know and what we need to know. Agric Syst 101(3):113–127
51. Wang W, Sun J, Liu C, Xue Z (2017) Application of immunostimulants in aquaculture: current knowledge and future perspectives. Aquac Res 48(1):1–23
52. Weindl I, Lotze-Campen H, Popp A, Müller C, Havlík P, Herrero M, Rolinski S et al (2015) Livestock in a changing climate: production system transitions as an adaptation strategy for agriculture. Environ Res Lett 10(9): 094021
53. Westhoek HJ, Rood GA, van den Berg M, Janse JH, Nijdam DS, Reudink MA, Stehfest EE (2011) The protein puzzle: the consumption and production of meat, dairy and fish in the European Union. Eur J Nutrit Food Safety 123–144

Chapter 9
Mitigation of Greenhouse Gases Emission and Low Carbon Technologies

9.1 Principles of Mitigation of Greenhouse Gases Emissions

The mitigation technologies for reducing greenhouse gas emissions from agriculture in general based on three basic principles. Those are (i) reduction of greenhouse gases produced in soil, (ii) retarding transformation and emissions of those gases from soil to atmosphere, and (iii) increase in consumption of GHGs into the soil-plant systems (Fig. 9.1). As we know, the net GHGs exchanges are the balance between production and consumption. Therefore, knowing the exact underlying mechanism and driver involved in the processes, we could effectively design and execute the mitigation technologies/practices for various production systems in different agro-ecological regions. In the following sections, we will discuss the mechanism and drivers of production and the principles of mitigation of major GHGs in agriculture, namely, methane, nitrous oxide, and carbon dioxide.

9.1.1 Mechanism and Drivers of Methane Production and its Mitigation Principles

Mechanism of methane production and emissions

The methane is primarily produced in wetlands and rice-paddy system. Its production and consumption are basically biologically mediated in wetlands and anthropogenic wetlands (submerged rice-paddy soil). Methane is emitted to the earth-atmosphere through diffusion, ebullition at the water-air interface, and through the capillary tube-like aerenchyma tissues in rice plants (Fig. 9.2). In the wetland, where plants do not have the passage through aerenchyma orientation, CH_4 mostly escaped from soil to atmosphere by diffusion or ebullition. Otherwise, it is trapped in the soil if undisturbed. In undisturbed rice soil, up to 85–90% of methane is emitted through

© The Editor(s) (if applicable) and The Author(s), under exclusive license to Springer Nature Singapore Pte Ltd. 2020

P. Bhattacharyya et al., *Climate Smart Agriculture*, Green Energy and Technology,
https://doi.org/10.1007/978-981-15-9132-7_9

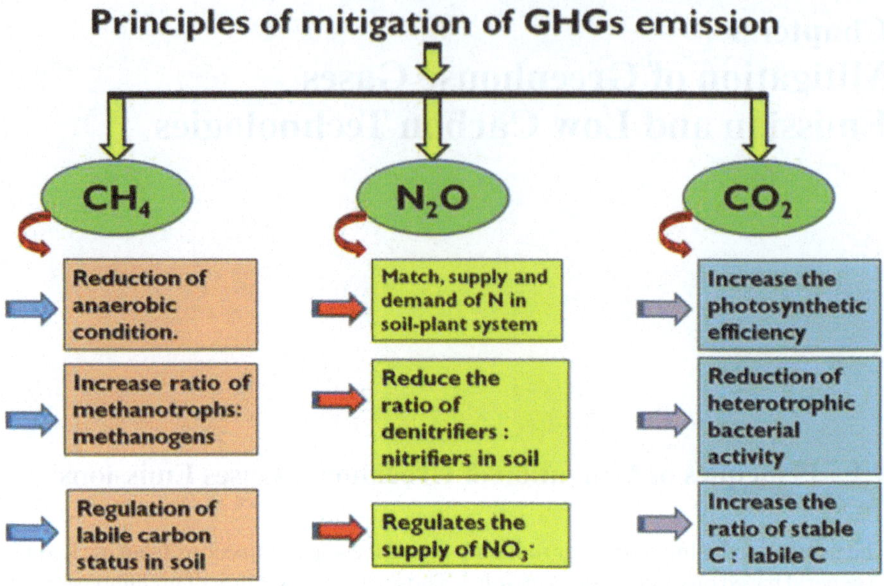

Fig. 9.1 Basic principles of mitigation of GHGs emission

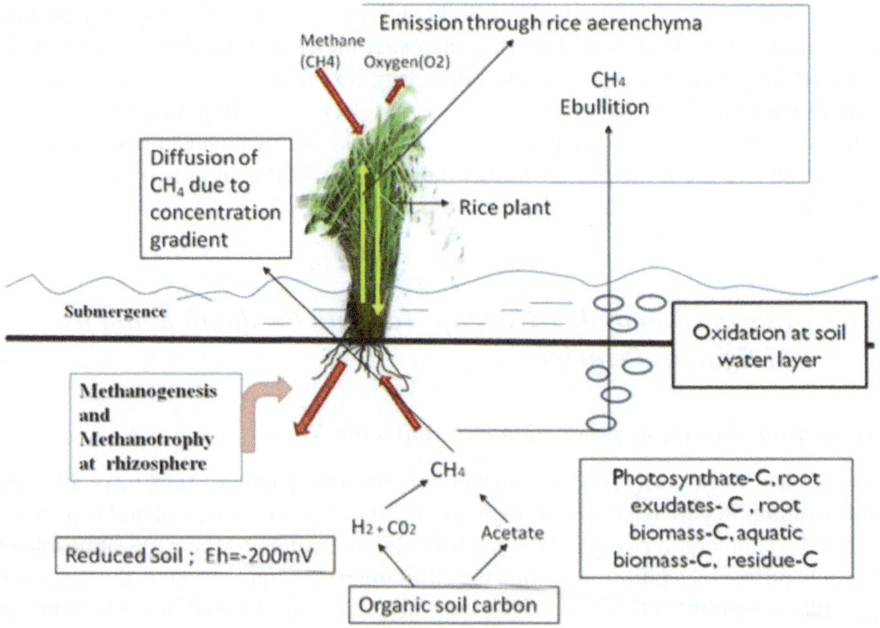

Fig. 9.2 Mechanism of methane emission from a typical rice field

Table 9.1 Different methanogens and the substrates preferred by them

Methanogens	Carbonaceous substrates preferred
Acetotrophs	Acetates
Hydrogenotrophs	Hydrogen and carbon dioxide
Formatotrophs	Formates
Methylotrophs	Methylated substrates
Alcoholotrophs	Alcohols

aerenchyma. Submerged rice soil facilitates the production of methane due to the presence of methanogenic bacteria that produces methane in anaerobic condition by utilizing easily decomposable organic carbon through methanogenesis. Rice plants also provide a conduit to emit methane from soil to atmosphere by its well organized capillary tube-like aerenchyma tissues that otherwise be used for transporting atmospheric oxygen to roots for respiration in submerged condition.

Now, specifically, we can say that the net emission of methane from the agricultural system is a resultant of methanogenesis (methane production) and methanotrophy (methane oxidation). The relative magnitude of both processes determine whether the net emission would be positive or negative. Broadly speaking, methanogens are methane-producing microorganisms and methanotrophs are methane-oxidizing bacteria. There are five major methanogen groups based on substrates utilization patterns found in agricultural soils (Table 9.1). Methanosaeta is less in number with respect to methanogenic population in soil but they contribute about two-thirds of total methane production in agriculture. Specifically, Acetotrophic methanogens have a slow growth rate. It is reported that about 70–80% of the methanogenic populations are utilized hydrogen and carbon dioxide but contributes only 25–30% of methane production in agriculture (Table 9.1).

The three primary things are required for the production of methane in the soil. One is the sufficient amount of carbonaceous substrate, second is a strictly anaerobic condition which maintains a redox potential around −200 mV and the presence of methanogenic bacteria. The type of major methanogenic bacteria and the substrates they utilized are presented in Table 9.1. Now, if you want to see the condition then according to submerged-chemistry oxygen reduces first, followed by nitrate, manganese, iron, sulfate, carbon dioxide, and hydrogen at definite redox potential (Table 9.2). In all those cases these substrates act as an electron donor. In Table 9.2, it can be clearly noticed that when reduction-potential would be around −200 mV or less, methane is produced in the soil [14].

Finally, when suitable carbon substrates are there and redox condition prevails, the methane is produced by methanogens and the process is known as methanogenesis. Simple carbon compounds like CO_2 and acetates contributes to major substrates for methanogenic fermentation.

About two-thirds of methane is produced in agriculture derived from the reduction of acetate (Eq. 9.1) and about one-third from reduction of carbon dioxide. In second case, the electrons are supplied by either hydrogen or formats (Eqs. 9.2 and

Table 9.2 Molecules with their reduced states and redox in submerged soil

Molecules/elements	Oxidized state	Reduced state	Redox potential (Eh) (millivolt) (mV)
Oxygen	O_2	H_2O	> (+) 100 mV
Nitrate	NO_3^-	N_2O	> (+) 200 mV
Manganese	Mn^{+4}	Mn^{+2}	(+) 200 to 0 mV
Iron	Fe^{+3}	Fe^{+2}	0 to (−) 150 mV
Sulfate	SO_4^{-2}	S^{-2}	< (−) 150 m V
Carbon dioxide	CO_2	CH_4	< (−) 200 mV

9.3). A small amount of methane is also generated by the oxidative-reductive dismutation of methanol or methylamine, specifically in marine sediments. Methanogenic microorganisms could also produce methane from secondary and cyclic alcohols or dimethyl sulfide. In all the cases carbon substrates act as an electron donor.

The major chemical reaction in methanogenesis:

$$CH_3COO^- + H^+ \rightarrow CH_4 + CO_2 \tag{9.1}$$

$$4H_2 + CO_2 \rightarrow CH_4 + 2H_2O \tag{9.2}$$

$$4HCOO^- + 4H^+ \rightarrow CH_4 + 3CO_2 + 2H_2O \tag{9.3}$$

The *Methanobacterium thermoautotrophicum* stains are mostly responsible for CO_2-reduction production pathways, whereas *Methanosarcina* and *Methanothrix* are responsible for the conversion of acetate to methane [11, 13, 15].

Methanotrophy is the oxidation of methane to CO_2 and H_2. In the transformation of the methyl group to CO_2, oxygen or other higher oxidation status molecules act as electron acceptors. Methanotrophs belong to a broad group of microorganisms that utilized 1–C (One carbon) compounds with C–C bond for their growth. These organisms are specifically able to use CH_4 for their metabolic activities. Methanotrophs are obligate aerobes that use molecular oxygen as the terminal electron acceptor. However, anaerobic CH_4 oxidation also noticed in SO_4^{-2} reducing environments [14]

Drivers of methane production and emission in agricultural soil

The major drivers of methane production and transmission in soil could be listed as soil redox potential, pH, dynamics of C substrates availability. Submerged rice soil acts as both source and sink of CH_4. However, the source and sink capacity depending on technological inputs and management practices followed. In rice production system CH_4 predominantly emits from soil to atmosphere through the aerenchyma pore spaces present in rice plants. So the orientation of aerenchyma tissue, pore spaces within the tissues at root-culm, and stem are the major driver of CH_4 transmission. Further, nutrient availability affects the construction and development of aerenchyma affect the emission process. The amount and nature of root exudation also play an important role in CH_4 production in soil. Other drivers are soil properties like structure, porosity, and hydraulic conductivity. Frequency and depth of flooding are also sensitive drivers of CH_4 emission in the rice-paddy system.

Principles of mitigation of CH_4 emission

Mitigation principles of methane emissions in agricultural soils are the following:

(i) Regulation of readily decomposable organic carbon in the soil.
(ii) Balancing labile carbon supply to methanogens from native soil as well as from external sources.
(iii) Controlling the processes of rhizodeposition and root exudations through rhizosphere modification.
(iv) Periodic maintenance of alternate wetting and drying in the soil to keep soil redox potential above −200 mV.
(v) Cutting down the continuous/prolonged submergence condition in rice by mid-season drainage.
(vi) Regulating ionic balance in flooded rice soil by keeping a considerable amount of Mn^{+2}, Fe^{+3}, SO_4^{-2} ions, so that redox potential would be around (−)50 to (−)150 mV or above. And those ions act as favorable electron donors and not allow acetate or CO_2 to reduce and produce methane.
(vii) Maintaining a relatively higher methanotrophic bacterial population over methanogens by practicing AWD (alternate wetting and drying) or aerobic rice and enriching the native methanotrophic population in the soil. So that relatively higher methane oxidation taken place in the soil as compared to methane production.

9.1.2 Mechanisms and Drivers of N_2O Production and Principles of Mitigation

Mechanism of N_2O production and emission

There is a trade-off mechanism between CH_4 and N_2O production in agricultural soil. Methane production requires continuous submergence but N_2O production needs a continuous cycle of aerobic and anaerobic conditions. In aerobic condition nitrate (NO_3^-) is produced from ammonium (NH_4^+) and $N_2O/NO/N_2$ is produced from NO_3^- in anaerobic condition. At oxidation reaction first nitrites (NO_2^-) are formed then nitrates (NO_3^-) by Nitrosomonas and Nitrobacteria species, respectively. While in the denitrification process, the NO_3^- is converted to either N_2O, NO, or N_2 depending on the redox condition of the soil by denitrifiers. Therefore, the conditions strictly favor CH_4 production (detailed discussed in an earlier section) not favorable for N_2O production and vice versa.

Rice fields are generally submerged; however, periodical drainage and irrigation is also a common feature in medium and upland ecologies. Therefore, both aerobic and anaerobic condition due to alternate wetting and drying cycles makes the soil system favorable for both oxidized and reduced nitrogen metabolism. Physical, chemical, and microbiological properties of rice/wetland soils influence the N-transformation processes and transmission of N_2O in different ways than that of observed in aerobic soil. Nitrous oxide is resultant of both nitrification and denitrification processes in soil. Nitrification is dominant in aerobic soil, whereas, denitrification dominants in submerged paddy or wetlands. Denitrification is the biochemical conversion of nitrate/nitrite to N_2 or N-oxides under anaerobic conditions (Eq. 9.4).

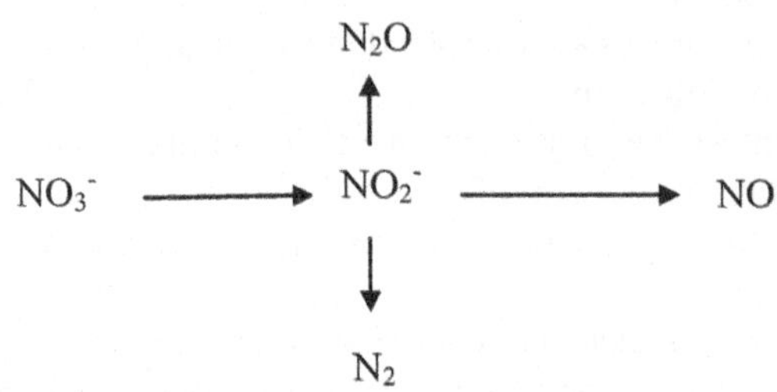

(9.4)

On the other hand, nitrification is the process of oxidation of ammonium to nitrite/nitrate and is equally responsible for the production of N_2O in the soil as it provides the necessary substrate for reduction. In these cases, a series of pathways were found via a few intermediate derivates like NH_2OH or NOH (Eq. 9.5) [39]

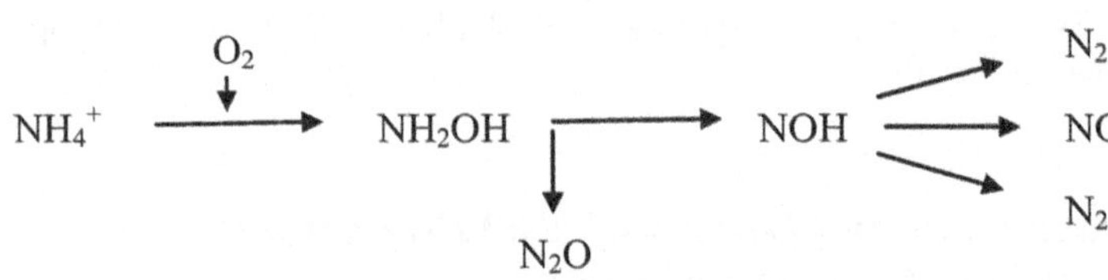

(9.5)

Drivers of N_2O production and emissions in agriculture

Several microbiological and physiological drivers affect N-transformation processes in the soil system and the subsequent emission of N_2O. The "hole-in-the-pipe" model nicely explains the drivers that control/regulate N_2O formation in soil [25] and emission from soil to the atmosphere. The simple pictorial model uses the analogy of a "leak pipe" of different holes and dimensions (Fig. 9.3) to describe the two distinct levels of control exist which regulate the transformation of NO, N_2O, N_2 in the soil-atmospheric system.

The pictorial/conceptual model could be explained by the following steps. At the first level, the rate of N cycling through a system that determines the amount of N_2 flowing through the pipe and influence the total transformation and emission. The second level is the drives, i.e., plant, water, soil, etc. that control the ratio of

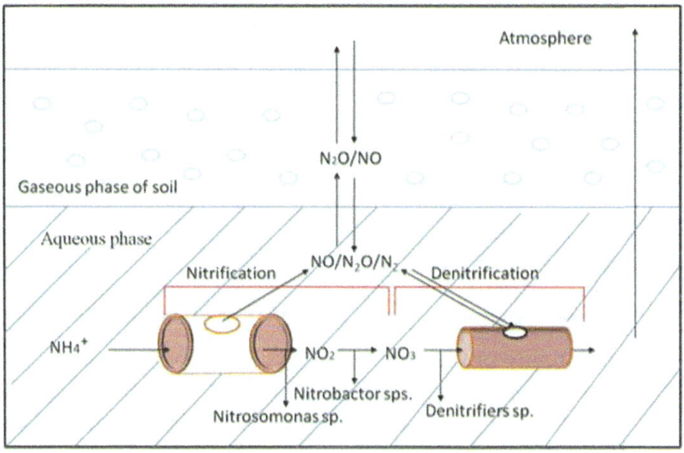

Fig. 9.3 Nitrous oxide production, transformation, and emission from soil to the atmosphere ("pipe-hole-theory"). *Source* Naqvi and Narontia [39]

N_2O: NO. The drivers are described as the leakage of pipe and their effect is represented by their relative hole size. For example, in well fertile soils "leaks" are more, so flows through the pipe are higher. The opposite is true for poor-fertile soil. In aerobic soils, where O_2 is more the nitrification "leaks" is higher and NO is dominant which is a relatively more oxidized form of N than N_2 and N_2O. In anaerobic (submerged/flooded) soil, when O_2 is limiting denitrification is the dominant process and higher N_2O production is evident. In extreme wet soil (continuous submerged soil) the denitrification process dominant for a long time and N_2O is further reduced to N_2. Apart from this, the type and amount of N-fertilizer used also regulates the N flows through the "pipe" and thereby influences the N_2O flux. Soil moisture status is the prime driver that regulates the transport of O_2 into the soil system and the fluxes of NO, N_2O, N_2 in the soil-atmosphere continuum. This driver is so important that the relative proportion of nitrification and denitrification and emission of N_2/NO/N_2O frequently described as a function of the water-filled porosity of the soil. Another important driver is the relative abundance of electron donors (soil organic carbon) and acceptors (O_2, NO_3^-, NO_2^-, SO_4^{-2}, etc.) present in the agricultural soil system. The rate of microbiologically mediated denitrification and production of N_2O in soil primarily depends on the availability of readily mineralizable organic carbon and the presence of water-soluble organic matter. Broadly, the addition of easily decomposable organic carbon enhances the N_2O production in soil having inherent NH_4^+ or applied with fertilizer NO_3^-. The soil texture, pH, and salinity (EC) significantly influence the transformation of N_2O by their direct and indirect effects on denitrification and nitrification processes. As for example, in light-textured soils, N_2O production could be increased up to 6 times as compared to heavy textured soil. In the loamy textured soils, N_2O production lasts for around 3 days only, whereas, in the silt textured soils it could go up to 10 days and in sandy soils, it is further extended up to 21 days. High salinity retards nitrification as well as denitrification, as N_2O reduction is susceptible to salt. Importantly, nitrification is sensitive to extreme pH. The optimum soil pH for nitrification is around 7–8. Soils with a considerable amount of NH_4^+ under aerobic situations showed significantly higher production of N_2O with increasing pH up to 8. However, above soil pH 8.2, nitrate accumulates in soil, and NO_3^- production is reduced as biological oxidation of nitrite by Nitrobacter is restricted.

Principles of mitigation of N_2O

The basic principles of reduction of N_2O emission from agriculture are followed:

(i) Matching the supply of nitrogen (from fertilizer/BNF/manure etc.) to the soil with the demand for nitrogen by crops. The matching should be site-specific and real time. So, that the N-use efficiency would be more and N loss will be less.
(ii) Control release of N from fertilizer to match the continuous demand of the crops. Slow-release fertilizer (coated urea) and nitrification inhibitors are the options.
(iii) Balancing the supply of other nutrients to crops to maximize the N-utilization.
(iv) Changing the method of fertilizer/manure application, so that N would be released slowly and no instant spike of N_2O flux happens.
(v) Regulating the cycle of oxidation-reduction in the soil so that the balance between nitrification and denitrification is maintained.
(vi) Controlling the availability of electron donors and acceptors responsible for N_2O production and transformation in soil.

9.1.3 Mechanism and Drivers of CO_2 Production in Agriculture and Principle of Mitigation

Mechanism of CO_2 production and emission

Photosynthesis and soil respiration are the primary two processes which regulate the terrestrial carbon balance in general and agricultural carbon budgeting in particular. Therefore, the net ecosystem carbon dioxide exchange is the net balance between gross primary production (GPP) and ecosystem respiration (RE). Gross primary production reflects the total photosynthetic assimilates, whereas, RE represents the respiratory CO_2 fluxes from roots (autotrophs), rhizosphere, soil, and microbial fauna (heterotrophs). Basal soil respiration is associated with soil organic carbon (SOC) decomposition which is the key pathway of carbon fluxes from agricultural soil to the atmosphere. It includes, microbial and roots respiration and decomposition of soil native organic matter. Soil CO_2 emission could be partitioned into basal soil CO_2 production, rhizospheric fluxes, soil microbial respiration, and root respiration with different turnover rate and residence time in the soil (Fig. 9.4).

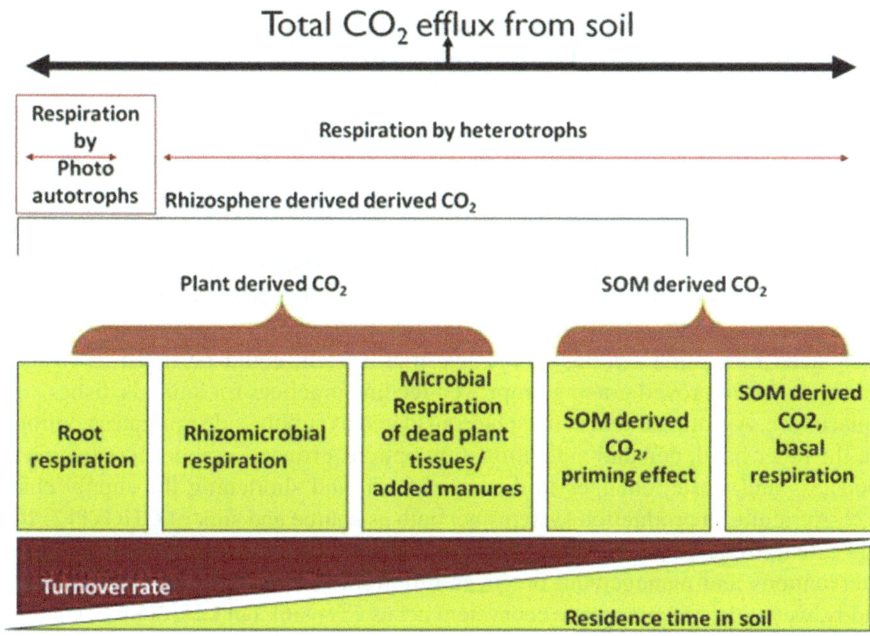

Fig. 9.4 The CO_2 flux-partitioning in the soil-plant system

Drivers of CO_2 production and emission

The drivers of CO_2 emission from soil to atmosphere are soil pH, temperature, and management practices like tillage, drainage, mode of fertilization, cropping sequence, type of crop, residue management, and amendments used. Apart from that puddling operation in rice-paddy is also played a significant role in CO_2 fluxes from agricultural soil.

Principle of mitigation of CO_2 emission

The principles of reduction of CO_2 fluxes from agricultural soils are listed below:

(i) Development of climate-smart cultivars through conventional and molecular breeding having a higher potential to harness atmospheric CO_2 both under ample and low light conditions.
(ii) Breeding of cultivars with higher GPP: RE ratio on a system basis and which also able to transfer stable form of carbon from aboveground portion to belowground parts in plant system.
(iii) Reducing the decomposition rate of SOM/SOC by any means.
(iv) Balanced application and utilization of easily decomposable organic carbon in the rhizosphere and or bulk soil so that heterotrophic activities could be controlled.
(v) Soil priming effects must be controlled and inhibited to enhance carbon sequestration and maintaining the old stock of stable carbon in the soil for a longer time.

9.2 Mitigation-Cum-Adaptation Options for Reducing GHGs Emissions from Agriculture

The 5th assessment reposts of IPCC (Intergovernmental Panel on Climate Change) categorized the mitigation options of GHGs emissions in agriculture into two distinct pathways, namely, supply-side and demand-side mitigation options. In a broader sense, supply-side mitigation options include the land-use changes both in crop and livestock and carbon sequestration in the soil-plant system. Cropping system manipulation, fertilizer management, biochar application, breeding of crops/plants with longer roots systems are the major options in the crop management side. On the other hand mitigation options in the livestock system could be listed as manure management, improved grazing, improved feeding practices for animals, fishes, and aquaculture. Agroforestry also must be considered as supply-side mitigation options. On the other hand, demand-side mitigation options primary includes a reduction in food loss and waste, changes in dietary pattern, and shortening the supply chain [52]. Agricultural production systems act both as source and sink of GHGs [12, 14]. The extent, frequency, and time at which it behaves differently depend on technical interventions and management practices followed. As, for example, lowland rice-paddy, wetlands, and mangrove ecosystem act as CO_2 sink but CH_4 source whereas, aerobic rice, pulse, and horticultural systems act as a CO_2 source. Crop varieties, balanced fertilization, water management, and amendment used in agriculture have definite and distinct roles in regulating and reducing GHGs emissions from agriculture. Some of them directly mitigate (reduce) the emissions and others if adopted increase the productivity and resilience of the system and at the same time reducing the emissions by increasing the resources use efficiency. The major mitigation and mitigation-cum-adaptation options are described in detail in the coming sections. The major mitigation and mitigation-cum-adaptation options for the reduction of CH_4, CO_2, and H_2O emission are presented as a flow chart in Fig. 9.5.

9.2.1 Supply-Side Mitigation Options

The total mitigation potential from supply-side options estimated as 2.0–2.5 Gt CO_2-eq yr^{-1} by 2030 [52]. Apart from this, the primary importance of supply-side mitigation options is that those could be directly applied to the production system and its stakeholders like farmers, retailers, and contribute significantly to livelihood and income generation. So, those options are relatively easily adaptable as effective mitigation practices hence broadly discussed in this section as compared to demand site options.

Effective use of fertilizer and manures

Deep placement of nitrogenous fertilizers particularly urea super granules reduces both CH_4 and N_2O emissions compared to urea broadcasting in rice [19]. Balanced

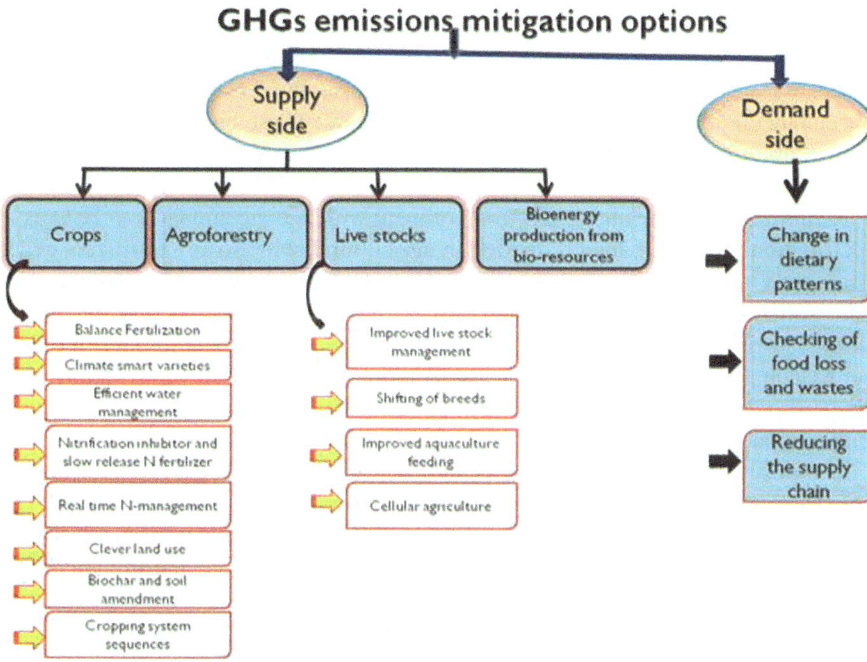

Fig. 9.5 Mitigation options of GHGs emissions

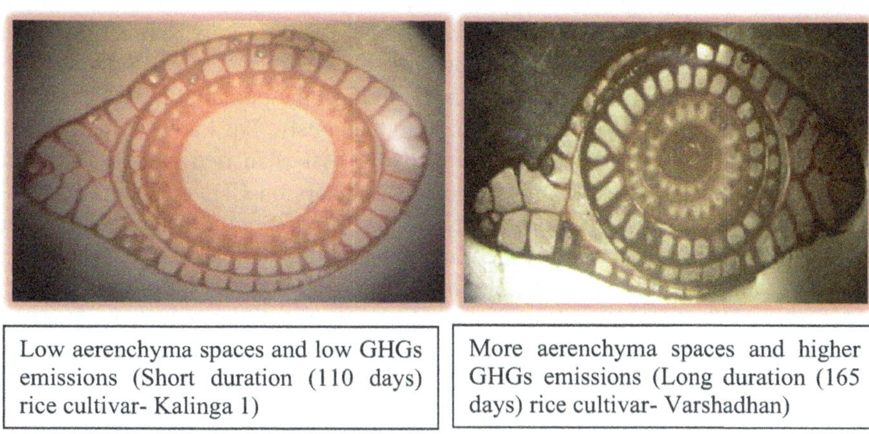

Fig. 9.6 Aerenchyma spaces orientation of two contrasting rice cultivars

application of phosphorus and potassium through sulfate-containing fertilizers considerably reduces the CH_4 emissions from rice by increasing fertilizer use efficiency as well as inhibiting CH_4 production in soil by the competitive effect of SO_4^{-2} ions. Sulfate ions compete with acetate and CO_2 for electron donors in submerged anaerobic soils. For example, the addition of potassium through potassium sulfate (K_2SO_4) significantly reduced CH_4 emission in rice production systems [53]. This is because adequate K in the rhizosphere increases the oxidizing power of roots and enhancing iron oxidation in the rice rhizosphere. These higher oxidative conditions inhibit the methanogenesis and thereby reduce the production and subsequent emission CH_4 in rice ecology.

As the sulfate, manganese and iron compete with acetate and $CO_2 + H_2O$ as an electron donor for methane production in agricultural soils, so any fertilizers generating those ions after dissolution in soil reduce CH_4 production. Therefore, ammonium sulfate [$NH_4(SO_4)_2$] as a source of N-fertilizer and ferrous and manganese sulfate (as micronutrient-fertilizer) are the good options for maintaining soil quality and reducing the methane emission in agricultural systems. However, ammonium sulfate should not be used in acidic soil, because there is a possibility of further soil acidification.

Compost with a low C: N ratio (C: N < 20: 1) compared to freshly prepared manure with a high C: N ratio is always recommended for reducing methane emission. Fresh organic manures tend to increase methanogenic activities in soil, particularly in a flooded condition. However, if we consider the system as a whole, the amount of CH_4 is also produced during making the compost. However, in relative terms considering the N-substitution made (from fertilizer) by compost, the GHGs emissions could be reduced from the systems as compared to only inorganic fertilizer application [16].

Introduction of climate-smart varieties

Previous two decades people are interested in climate-resilient cultivars that can cope up with climate change vagaries like drought, flood, salinity, submergence, etc [8, 9]. The approaches shifted to climate-smart varieties recently which not only have the resilience to climatic vagaries but also have the potential to reduce/curtail down GHGs (CH_4 and N_2O) emission, simultaneously. We know about 90% of CH_4 emitted through well-oriented aerenchyma tissues in rice plants. There are varietal differences that exist among the rice cultivars on CH_4 emission potential [10, 48]. Therefore, suitable ecology-based selection of rice cultivars could help to mitigate CH_4 emission in rice production systems. For example, short duration rice cultivars (Kalinga 1) have higher CH_4 mitigation potential than long-duration ones (Varshadhan) (Fig. 9.6) [10]. Moreover, newly developed climate-smart rice varieties (drought and submergence tolerant), like CR Dhan 801, and CR Dhan 800 could be chosen as mitigation-cum-adaptation options to GHGs emissions. The introduction of submergence tolerance rice cultivar (Swarna Sub1) also bears the potential to reduce the CH_4 and N_2O emissions in lowland ecologies. Terminal heat-tolerant wheat cultivar P159-2/P157-8 and short duration pulse variety (cv. Virat, for green gram) have the potential to reduce N_2O and CO_2 emission, respectively, due to less nitrogen and water requirement and very short duration (less rhizosphere activity).

Efficient water management

Water management plays a crucial role in GHGs production, transformation, and emission in agricultural systems. Alternate wetting and drying in direct seeded rice

(DSR) has been recommended to curtail down CH_4 emission of about 20–30% as compared to transplanted rice-paddy. Considering the emerging crisis of water and labor in rice cultivation, this technology has a huge potential for mitigation-cum-adaptation of GHGs emissions in south-Asian rice-growing countries. Mid-season drainage between maximum tillering and panicle initiation stage of transplanted rice-paddy has the potential to reduce the methane emission up to 35–50%. These management practices also have good adaptation potential. Matching of irrigation scheduling with N-fertilization was found a very effective option for the reduction of N_2O emission in rice, wheat, maize, and sugarcane. Drip and sprinkler irrigation systems in horticultural crops, particularly in orchards have been identified as a mitigation option of N_2O and CO_2 emissions. Increasing the number of split irrigation in wheat in sandy or sandy loam soil has the potential to increase N-use efficiency and reduction of losses of N through N_2O. Foliar spray of nitrogen in wetland crops could be an effective option for the reduction of both N_2O and CH_4 emission in those ecosystems.

Use of nitrification inhibitors and slow-release nitrogenous fertilizer

Nitrification inhibitor is a well-known mitigation option for N_2O emission which restricts the first step of nitrification, i.e., conversion of NH_4^+ (ammonium ion) to NO_3^{-1} (nitrate ion). Two organic nitrification inhibitors, namely, Nmin and Karanj oil were found effective in flooded alluvial (Inceptisols) rice soil. The Nmin (having 5% of Neem bitter) acts as a nitrification inhibitor which is the concentrated extract of neem (*Azadirachta indica*) seed. Another widely used nitrification inhibitor in rice and wheat is Dicynamide (DCD). This is a bacteriostatic inhibitor that acts very specifically on *Nitrosomonas* sps.

Slow-release nitrogenous fertilizer helps to match the nutrient demand of crop to the nutrient supply from soil/fertilizer. At the same time, it reduces the production and emission of N_2O from soil to the atmosphere. The coating of urea with neem and sulfur significantly reduced the N_2O emission (5–10%) in rice along with increasing NUE by 10% [34]. Urea super granules or urea briquettes placed in the reduced zone of soil is a good option for the reduction of N_2O (20–25%) emission in lowland rice [19]. Coated urea briquettes with fly ash would further increase the potential of enhancing NUE and reduction of N_2O emission in the submerged rice-paddy system.

Judicious real-time N-management

Real-time N-management refers to the matching of the supply of N from soil system to the demand of crop-system. There are various approaches followed to real-time nutrient management like the application of fertilizer at critical growth stages, use of leaf color chart for N, use of SPAD meter, and sensor-based technology for identifying the nutrient deficiency for fertilizer application [36]. These approaches basically came from the purpose of increasing fertilizer use efficiencies in general and nitrogen use efficiency in particular. Those have the capacity to increase NUE of present-day 20–30% in agriculture to around 40–45%. As these approaches reduce the losses and facilitate use efficiency, it indirectly facilitates the reduction of emission of N_2O and CO_2. Real-time N-management is the need of the day and has the considerable potential for adaptation in different crops (rice, wheat, maize, sugarcane, horticulture) and different ecologies (low land, upland, medium land, and hilly regions). Recently, crop-specific leaf color chart has also been developed for N-management like CLCC (Customized leaf color chart) of rice, wheat, and maize. The LCC is an easy, cost-effective, and adaptable technology and has the potential to reduce N_2O emission by 5–10% [7].

Addition of biochar and soil amendment

Biochar not only adds recalcitrant carbon pools but also improves the physical properties (porosity) of the soil, hence, increasing agricultural productivity. It has also climate-mitigation potential, stems primarily from its highly recalcitrant nature, which slows the rate at which photosynthetically fixed carbon in soil is returned to the atmosphere [32]. Soil amendments are any organic or inorganic additive added to soil to improve its physical or chemical properties. Unlike fertilizers, the exact ingredients and chemical composition of soil amendments vary among different sources. Soil amendments can be used to rebuild soils that have been damaged by improper management [3]. A wide variety of soil amendments (geological or plant-derived materials) can be used to mitigate GHGs emission in agriculture which are alternative nutrient sources for the crop. As for example, phosphogypsum is used as a soil amendment. It supplements calcium and sulfur to crops. The high content of sulfate in phosphogypsum prevents methane formation as well as CH_4 emissions due to stronger competition for substrates (hydrogen or acetate) among methanogens and sulfur-reducing bacteria [30]. Basic slag is another industrial by-product of the steel industry, very effectively used as a soil amendment in agriculture. It contains a high amount of available silicate, free iron and manganese oxides, which act as electron acceptors, and an oxidizing agent (methane oxidation) in rice farming [4, 5]. Fly ash, a by-product of the coal industry, improves the physical properties of soil like soil structure, lower bulk density, higher water-holding capacity, etc. It also enhances the carbon sequestration in the soil. An experimental study revealed that 1 ton of fly ash could sequester up to 26 kg of CO_2, which confirmed the possibility to use fly ash for CO_2 mitigation.

Clever land-use management

Converting wetland for aerobic crops significantly enhances the CH_4 emission at a spark. However, this conversion never is encouraged unless it is essential; otherwise, wetland should be restored not only for reducing CH_4 emission but also storing a huge amount of carbon (sink of CO_2). We should practice clever land-use management by discouraging conversion of wetland to aerobic crops or aquaculture or conversion of mangrove to rice-aquaculture systems.

Another important aspect is how we manage the rice-paddy during summer and winter that would influence the CH_4 emission during the subsequent wet season. Intelligent carbon and nitrogen management in summer, winter, and fellows have the potential to reduce GHG emission in rice-paddy during the wet season by regulating soil labile carbon, and the population of methanotrophic, methanogenic, and denitrifying bacteria.

Cropping system and cropping sequence manipulation

Cropping system, cropping sequence, and cultural practices significantly influence the GHGs emissions in the agricultural production system. As for example, the rice-pulse system has 30–35% GHG mitigation potential as compared to the rice-rice system [41]. Direct seeding of rice helps in the reduction of both CH_4 and N_2O than the transplanted rice-paddy system [43]. Zero tillage favors carbon build-up and reduction of global warming potential (GWP) compared to the conventional rice-wheat system (84–88%) [43]. Similarly, zero and minimum tillage showed the mitigation potential of CH_4 and N_2O in rice-wheat cropping systems. Present-day, the use of second-generation machinery and *in situ* residue management has been showing significant potential to reduce GHGs emissions in rice-wheat and wheat-maize cropping systems.

Improved livestock management and shifting of breeds

Improve livestock management for mitigation of GHG emission mostly focused on the reduction in enteric methane production, reducing N_2O emission through manure management and land-use management for pasture for sequestering carbon. Livestock management options have the potential to reduce GHG emission in the tune of 1.8–2.4 Gt CO_2-eq yr^{-1} [24, 27]. Feeding better quality diets (includes industrially produced microbial protein, legumes, improved pasture sps., etc.) reduce the quantity of GHGs emission per unit animal product [45]. Improved feeding system, a dairy-crop mixed system for dual-purpose (milk, crop, and meat), manipulation of rumen microflora, promoting lower methane-producing breeds, and use of feed additives are very effective options for reducing GHG emission from livestock sectors [28]. Controlling of low productive breeds (low milk producing but higher CH_4 generator) and replacing them with better-feed crossbreed animals with improving grazing management both in developed and developing countries have the vast potential to curb GHGs emissions and sustaining the supply of livestock products [1]. Environment favorable adoption of breeds, shifting from ruminants to monogastric (e.g., from beef to pig or poultry) are the other eco-friendly options for curbing GHGs emissions from the livestock sector.

Manure management requires a systematic approach starting from feeding a balanced diet (so that excretion contains less labile nitrogen) to livestock, collection of excreta/bedding material, storing those including slurry to anaerobic tanks (to reduce CH_4 and N_2O emission) and delivery to the field. In each step, there is a potential to reduce N_2O as well as CH_4 emissions [18], [17]. Methanogenesis in ruminants stomachs could be reduced by supplementing the diet with dietary lipid or antimethanogenic agents like algae (*Asparogopsis taxiformis*), 3-nitro-oxypropanol (3-NOP). And with those methane emissions from ruminants could be reduced in the tune of 30–75% [29].

GHGs mitigation in aquaculture and cellular agriculture

Several options exist to reduce CO_2 emission from aquaculture (up to 21% per ton of fish production). However, options must be integrated according to site demand and economic feasibility. The primary options are replacement of fish-based ingredients with crop-based ingredients; improved fish health and feed conversion rates; use of renewable energy; and increase in input-use efficiency [6].

Cellular agriculture is the future of the livestock science/industry which includes cultured meat, in vitro meat, synthetic meat, hydroponic meat, industrial production of milk, egg white, and leather [49]. It has a tremendous mitigation potential with respect to reducing land use, ruminant CH_4 production, and manures generated GHGs [31]. However, most of these technologies are still not commercialized and their actual impact on the environment is not yet tested.

Reduction of GHGs emission through Agroforestry

Agroforestry is listed separately as it has an enormous and distinct potential to curb the emission of CH_4, N_2O, CO_2 in the agricultural system (both crop and livestock). It is equally important in both developing and developed countries. Apart from its carbon sequestration potential, an additional benefit of this system includes organic matter build-up in soil with higher moisture holding capacity [14]. Soil degradation and off-site carbon loss through erosion could also be checked by this system. However, carbon sequestration potential and greenhouse gas mitigation potential of agroforestry system are widely varied with climatic regions (humid > temperate > semi-arid) with an estimated carbon sequestration potential of 1.5–3.5 Mg C ha^{-1} yr^{-1} [38].

Combined options by integrating crop and livestock to reduce GHGs emission

Crop and livestock production systems are deeply interdependent. There is continuous energy flow from one system to another and vice versa. Therefore, combined options which could address both the sectors, energy flow between them, by-product utilization and marketing would give a better possibility to reduce GHG emission from agriculture. Options like feeding livestock on crop waste and use unproductive land (not suitable for efficient crop production.) for pasture, producing microbial protein from sewage streams, and feeding those to livestock in place of soybean protein could reduce the emission by 7% [45]. So called bio-waste (lingo-cellulosic crop residue, waste stream from manure, food waste, etc.) can be converted to generate CH_4 (bio-energy) and microbial protein [21, 45]. Second-generation biofuels from lingo-cellulosic crop residue have an enormous potential to curb GHGs emissions both in direct and indirect ways. In this aspects production of bio-diesel, bio-pyrolysis, production of cellulosic bio-ethanol are the potential options available which not only reduce the GHG emission but at the same time supplement a lot of the energy requirement on a sustainable basis [42].

However, Paustian et al. [44] demonstrated that most of the individual mitigation options in crop/livestock management would have a little effect per unit of land, and proposed multiple options should be aggregated for crop and grazing land. For example, a combination of reducing tillage, balanced nutrient/fertilizer management, and improved crop rotation have a higher mitigation potential than any of the single intervention. Another important point to be remembered, individual and integrated mitigation options should also be judged based on GHG life cycle analysis, considering their co-benefits (as ecosystem services) and region-specific full-cost carbon accounting [44].

9.2.2 *Demand-Side Mitigation Options*

Demand-side mitigation options are primarily driven by socio-economic factors like population growth, global food demand, and policy fluctuation due to political turmoil, religion, and cast preferences. Therefore, even there would be technical potential or environmental concern, those options are difficult to adapt and promote. However, there is potential to mitigate in the tune of 2.7–5.7 Gt CO_2-eq yr^{-1} by the year 2050, based on the preference of diet and region. Demand site mitigation options primarily include food choice (diet preference), food loss and waste, and length of the supply chain (producer to consumer).

Change in dietary pattern

In general GHGs emission is more in the non-vegetarian diet than a vegetarian diet. Further, the non-vegetarian diet having "ruminant meat" contributes maximum to GHGs emissions to the atmosphere. Broadly speaking, the higher consumption of animal-based foods leads to higher environmental impact, while plant-based food has a lower environmental impact [40]. More specifically, ruminant meat (beef and lamb) has identified single food with a high negative impact on the environment on a global basis. A comparative study showed that flexitarian (limited meat + dairy) has the highest potential to mitigate GHG emission followed by "healthy diet" (limited sugar + meat + dairy), and vegetarian over non-vegetarian (ruminant meat consumer) [27]. In a healthy diet scenario, it was assumed that protein requirement would be fully compensated by a higher intake of pulses. This reduction-potential estimation includes all the aspects of land-use change, area conversion, ruminant CH_4 reduction, production process-CO_2 emission. However, Smith et al. [50, 51] recommended the diet scenario analysis should be done as a "per day per capita calorie intake" basis.

Checking and regulating food loss and waste

Robust life cycle analysis for around 50 years (1961–2011) showed that global food loss and waste contributed around 10% of total anthropogenic GHG emission (~44 Gt CO_2-eq yr^{-1}) [46]. In general, a large quantity of food is lost in developing countries due to poor storage, mishandling, and poor infrastructure, whereas a large portion of food is wasted in developed countries [26] due to negligence or narrow preferences. However, there is vast potential exist in changing consumer behavior (to reduce per capita over-consumption) and infrastructure development for the reduction of emissions related to food loss and waste. Around 20% of food production was wasted in Europe and North America and 30% of food production was wasted in Africa (during 2007) [22]. Specifically, 3% of world N_2O emission is associated with loss and waste of sheep meat, poultry meat, milk, pig meat, and potatoes in 2009

[47]. However, Alexander et al. [2] pointed out the largest absolute losses of food occurs as crop-biomass before harvest. Even losses from the harvested crop are also substantial, about 44% of crop dry matter is lost globally before human consumption [2].

If we critically analyze, then we can see most of the food is lost during production, post-harvest handling, processing, and transportation in developing countries (low-income countries). The main problem associated with that is poor infrastructure including lack of cold storage, poor processing, and packaging facilities and real-time transportation of produce to the market. These problems could be addressed by financial assistance and a robust policy-framework. So there is a good opportunity to minimize GHG emissions in developing countries by improving processing and package industries, expansion of road and infrastructure, and making effective cold chain systems from production site to the market place [22]. On the other side, in developed countries foods are wasted that could be reduced by changing consumer perception and utilizing packaging and processing technologies more effectively [23, 54] and thereby mitigating GHGs emissions to a considerate amount. There is potential to reduce N_2O emission by 0.3 Gt CO_2-eq yr^{-1} by avoiding complete loss and waste of food in the agricultural sector [47]. Avoiding food loss and waste is very often considered as a climate change adaptation option for decreasing yield due to climate vagaries, however, it could very well be considered as a mitigation option for reducing GHG emission in the light of reducing agricultural expansion for producing excess food.

Regulating the supply chain

Local production based on environmental suitability and consumption at the local level is the key to this mitigation option. It not only minimizes the losses due to transportation at the same time ensure food security and reduces GHG emission associated with food loss. But we should be careful; the environmental benefit with local food production-consumption may be offset, if an inefficient production system is followed. For example, use local livestock products from unproductive breed may reduce GHG emission by reducing the supply chain but sometimes it is less beneficial to use imported animal products from the highly productive system with has much lesser GHG footprint. Therefore, the supply chain should be regulated based on a life cycle-based GHGs footprint analysis of the whole system basis rather than an individual component.

9.3 Strategies and Management Practices to Mitigate GHGs Emission

In the previous sections, we discussed mostly principles and overall mitigation and or mitigation-cum-adaptation options of GHGs emissions in agriculture. In this section, we will discuss the specific strategies and management practices to reduce GHGs emissions with some field-level examples. However, primarily we will concentrate on supply-side mitigation strategies that are technology-driven. Demand-side mitigation strategies are not discussed in detail as those are mostly socio-economic and policy-driven.

9.3.1 Strategies and Management Practices to Mitigate Methane Emission

The specific strategies for reducing CH_4 emission from agriculture in general and rice, in particular, could be bulleted as,

(i) Drainage for 2–3 days in transplanted rice-paddy in between maximum tillering and panicle initiation stages.
(ii) Intermittent irrigation in irrigated medium and lowland rice; and alternate wetting-drying cycle for *rainfed* rice.
(iii) Maintaining sulfate in a reduced zone of acid sulfate soil to keep CH_4 emission from the system at a lower level.
(iv) Replacing conventional puddled rice belt to direct-seeded rice (DSR) belt with proper weed and water management. New generation selective herbicides and second-generation farm machinery must be integrated with DSR for getting sustainable yield with lower CH_4 emission.
(v) Dry-direct seeded rice in place of wet-direct seeded rice in medium to upland ecologies.
(vi) Transplanting of relatively older seedling (21–30 days) of rice instead of very young seedling (8–10 days) in transplanted rice.
(vii) Selection and introduction of low CH_4 emitting rice cultivars in suitable ecology.
(viii) Aerobic decomposition or rice-straw is advisable during winter/summer in South-Asian countries to reduce CH_4 emissions in the subsequent wet season.
(ix) *In situ* green manuring as paired row cropping (rice-rice-*Sesbania)* and at proper time incorporation (25–30 day after sowing) instead of traditional green manuring (growing of *Sesbania aculeata* for 40–45 days and then incorporation during puddling) in rice.
(x) The proper time of planting also reduces the emissions of CH_4 as it is influenced by diurnal temperature variations. Methane emissions were found more in late-planted rice during the wet season.
(xi) Composted rice-straw application instead of fresh (uncomposted) one. It regulates the labile carbon supply to methanogens.
(xii) Application of sulfate fertilizer in any form to reduce the CH_4 production and subsequent emission.
(xiii) High-frequency low amount irrigation scheduling to rice and all other wetland crops.

(xiv) Applications of soil amendments like biochar @ 5 t ha^{-1} (once in three years); phosphogypsum (2 t ha^{-1}) and basic slag (1 t ha^{-1} in the rice-paddy system).
(xv) The addition of butachlor, hexachlorocyclohexane, carbofuran with proper dose has shown their potential to reduce CH_4 emission in agriculture. Those molecules either hinder the dropping of redox potential in the soil to −200 mV or inhibit the activities of methanogenic bacteria.

9.3.2 Strategies and Management Practices for the Reduction of N_2O Emission

We have discussed in previous sections (Sects. 9.1 and 9.2) that there is a direct relationship between nitrogen use efficiency (NUE) and N_2O emission in the agricultural system. Strategies used for increasing NUE indirectly help in the reduction of losses of nitrogen in general and N_2O emission in particular. Therefore, both strategies for enhancing NUE and or reducing N_2O emissions mostly concentrated on the demand-supply chain of N, type, and method of N-fertilizer application and nitrification inhibitors. Some important strategies are listed below:

(i) Two to three split applications of nitrogenous fertilizers based on either critical physiological stage of crop or using leaf color chart (LCC; CLCC for rice), SPAD meter, etc [37].
(ii) Site-specific nutrient management takes into consideration the nutrient supply capacity of soil and yield targets.
(iii) Deep placement of urea in the reduced zone in rice-paddy helps in the reduction of N_2O emission.
(iv) Application of coated ammonium nitrate that could reduce N_2O emission in arable soil as compare to uncoated ammonium sulfate by 3–7 folds.
(v) Nitrification inhibitor like dicyandiamide, wax-coated calcium carbide, neem cake, neem oil, Karanja seed extract also some of the plant-derived nitrification inhibitors effectively reduce N_2O emission in rice-wheat cropping systems. Dicyandiamide (DCD) is a widely used inhibitor for N_2O emission [35].
(vi) Sprinkler and drip irrigation reduces N_2O emission as compared to furrow irrigation in cereals as well as horticultural crops by increasing NUE. On the contrary continuous submergence also reduces N_2O emission as compared to furrow or intermittent irrigation to crop by providing the continuous anaerobic condition, detrimental for nitrification and subsequent N_2O emission.
(vii) Zero and minimum tillage reduces the N_2O emission compared to conventional tillage in the rice-wheat system by increasing the compactness of soil and thereby reducing the activities of nitrifiers and denitrifiers.
(viii) Application of biochar reduces the N_2O emission in the rice-wheat and maize system by increasing the ratio of recalcitrant carbon to labile carbon and thereby reducing the process of mineralization and nitrification.

9.3.3 Strategies and Management Practices for Reducing CO_2 Emissions in Agriculture

The agricultural system frequently described as a CO_2 neutral system or a major sink of CO_2 from the atmosphere. In true sense, this system captured a large amount of CO_2 as the raw material for photosynthesis, the primary and most important process of the earth-atmosphere system. However, plant and soil also respire and emit CO_2. If we do not consider the net balance of ecosystem CO_2 exchanges, then we can select some strategies which could decrease the CO_2 emission from soil to atmosphere and make the ecological CO_2 exchanges more negative (i.e. higher uptake of CO_2 than emitted in the soil-plant system). Few of these strategies are listed below:

(i) Minimum tillage or zero tillage reduces the CO_2 emission in the rice-wheat cropping system compared to conventional tillage. The relative compactness developed due to reduced tillage reduces the activities of heterotrophs as compared to conventional tillage, where soil aggregate is broken and the soil becomes pulverized. Soil aggregation also protects the aggregate trapped carbon (particulate organic carbon in aggregate) to decomposition, hence CO_2 emission would be less.

(ii) Continuous submergence in the rice-paddy system although facilitates CH_4 production through the anaerobic decomposition of soil organic carbon (being less efficient than aerobic decomposition) but reduces CO_2 emission from the system.

(iii) The introduction of legume in rice-based cropping (rice-maize-legume) reduced the CO_2 emission compared to maize cultivation alone.

(iv) Biochar and mulch (in sequence) addition in the agricultural and or horticultural system have the huge potential to reduce CO_2 emission and increase C-sequestration by adding stable carbon pool in soil and regulating labile C fractions and soil respiration.

9.4 Mitigation Potential of Important Low Carbon Technologies in the Rice-wheat System

All the mitigation options and strategies described in previous sections could be classified as low carbon technologies. In this section, we would specifically summarize some of the important and adapted low carbon technologies in the rice-wheat system which covered almost 50% of the world's agriculture. The different technologies with their mitigation potential are presented in Table 9.3.

Table 9.3 Low carbon technologies and their GHGs mitigation potential

Low carbon technologies	% reduction in GHGs emission or GWP	% increase of GHG emission or GWP	Reference
Replacing rice-maize cropping system to rice-pulse	35% CO_2	–	Neogi et al. [41]
Mid season drainage in rice (between MT-PI stage) over continuous submergence	30–35% CH_4	5–10% N_2O	Lu et al. [33]
Incorporation of rice-straw in lowland rice-rice system	10% N_2O	8% CH_4	Bhattacharyya et al. [16]
Real time N-management through leaf color chart to rice-wheat system	11% N_2O	–	Bhatia et al. [7]
Direct seeding of rice (DSR) over conventional rice-wheat cropping system	70–75% CH_4	5–10% N_2O	Pathak and Aggarwal [43]
Aerobic rice over conventional transplanted rice	21–41% GWP	–	Pathak and Aggarwal [43]
Zero tillage wheat over conventional rice-wheat	84–88% less GWP	–	Pathak and Aggarwal [43]
Nitrification inhibitor over N-fertilizer through urea	7–16% less GWP	–	Pathak and Aggarwal [43]
Introduction of low GHG emitting cultivar over conventional HYV	30–40% CH_4	–	Bhattacharyya et al. [10]
Paired row (rice + *Sesbania*) in DSR	8–10% CH_4 3–5% GWP		Dash et al. [20]

References

1. ADAS Ltd, UK (2015) Study to model the impact of controlling endemic cattle diseases and conditions on national cattle productivity, agricultural performance and greenhouse gas emissions. https://sciencesearch.defra.gov.uk/Document.aspx?Document=13320AC0120_Finalreport
2. Alexander P, Brown C, Arneth A, Finnigan J, Moran D, Rounsevell MDAA (2017) Losses, inefficiencies and waste in the global food system. Agric Syst 153:190–200. https://doi.org/10.1016/j.agsy.2017.01.014

3. Ali MA, Farouque G, Haque M, Kabir A (2012) Influence of soil amendments on mitigating methane emissions and sustaining rice productivity in paddy soil ecosystems of Bangladesh. J Environ Sci Nat Resour 5:179–185
4. Ali MA, Hoque MA, Kim PJ (2013) Mitigating global warming potentials of methane and nitrous oxide gases from rice paddies under different irrigation regimes. AMBIO 42:357–358
5. Ali MA, Lee CH, Kim PJ (2008) Effect of silicate fertilizer on reducing methane emission during rice cultivation. Biol Fertil Soils 44:597–604
6. Barange M, Bahri T, Beveridge MCM, Cochrane KL, Funge-Smith S, Poulain F (2018) Impacts of climate change on fisheries and aquaculture: synthesis of current knowledge, adaptation and mitigation options. FAO Fisheries and Aquaculture Technical Paper No. 627. Rome, Italy, 628 pp
7. Bhatia A, Pathak H, Jain N, Singh PK, Tomer R (2012) Greenhouse gas mitigation in rice–wheat system with leaf color chart-based urea application. Environ Monit Assess 184(5):3095–3107
8. Bhattacharyya P, Barman D (2018) Crop residue management and greenhouse gases emissions in tropical rice lands. In: Soil management and climate change. Academic Press, pp 323–335
9. Bhattacharyya P, Pathak H (2017) Climate resilient agriculture: impacts and options for adaptation in India. SATSA Mukhaptra Ann Tech Issue 21:35–45
10. Bhattacharyya P, Dash PK, Swain CK, Padhy SR, Roy KS, Neogi S, Berliner J, Adak T, Pokhare SS, Baig MJ, Mohapatra T (2019) Mechanism of plant mediated methane emission in tropical lowland rice. Sci Total Environ 651:84–92
11. Bhattacharyya P, Nayak AK, Neogi S, Roy KS (2013) Greenhouse gas emission reduction strategies in agriculture to mitigate climate change. Central Research Institute for Dryland Agriculture, Santoshnagar, Hyderabad
12. Bhattacharyya P, Neogi S, Roy KS, Dash PK, Nayak AK, Mohapatra T (2014) Tropical low land rice ecosystem is a net carbon sink. Agr Ecosyst Environ 189:127–135
13. Bhattacharyya P, Roy KS, Neogi S (2017) Changes in soil–plant–microbes interactions in anticipated climatic change conditions. In: Adaptive soil management: from theory to practices. Springer, Singapore, pp 261–275
14. Bhattacharyya P, Roy KS, Nayak AK (2016) Greenhouse gas emission from agriculture: monitoring, quantification and mitigation. Narendra Publishing House, New Delhi. ISBN 13-9789384337964
15. Bhattacharyya P, Roy KS, Nayak AK, Shahid M, Lal B, Gautam P, Mohapatra T (2017) Metagenomic assessment of methane production-oxidation and nitrogen metabolism of long term manured systems in lowland rice paddy. Sci Total Environ 586:1245–1253
16. Bhattacharyya P, Roy KS, Neogi S, Adhya TK, Rao KS, Manna MC (2012) Effects of rice straw and nitrogen fertilization on greenhouse gas emissions and carbon storage in tropical flooded soil planted with rice. Soil and Tillage Res 124:119–130
17. Bourdin F, Sakrabani R, Kibblewhite MG, Lanigan GJ (2014) Effect of slurry dry matter content, application technique and timing on emissions of ammonia and greenhouse gas from cattle slurry applied to grassland soils in Ireland. Agric Ecosyst Environ 188:122–133. https://doi.org/10.1016/j.agee.2014.02.025
18. Bouwman L, Coauthors (2013) Exploring global changes in nitrogen and phos-phorus cycles in agriculture induced by livestock production over. Proc Natl Acad Sci 110:20882–20887. https://doi.org/10.1073/pnas.1206191109
19. Chatterjee D, Mohanty S, Guru PK, Swain CK, Tripathi R, Shahid M, Kumar U, Kumar A, Bhattacharyya P, Gautam P, Lal B (2018) Comparative assessment of urea briquette applicators on greenhouse gas emission, nitrogen loss and soil enzymatic activities in tropical lowland rice. Agr Ecosyst Environ 252:178–190
20. Dash PK, Bhattacharyya P, Shahid M, Roy KS, Swain CK, Tripathi R, Nayak AK (2017) Low carbon resource conservation techniques for energy savings, carbon gain and lowering GHGs emission in lowland transplanted rice. Soil and Tillage Research 174:45–57
21. De Clercq D, Wen Z, Fan F, Caicedo L (2016) Biomethane production potential from restaurant food waste in megacities and project level-bottlenecks: a case study in Beijing. Renew Sustain Energy Rev 59:1676–1685. https://doi.org/10.1016/J.RSER.2015.12.323.10.1038/nclimate1458

22. FAO (2011) Global food losses and food waste: extent, causes and prevention. FAO, Rome
23. FAO (2013) Food wastage footprint. Impacts on natural resources, 63 pp
24. FAO (2013). The state of food and agriculture. Rome, 114 pp. https://www.fao.org/docrep/018/i3300e/i3300e.pdf
25. Firestone MK, Davidson EA (1989) Microbiological basis of NO and N_2O production and consumption in soil. In: Andreae MO, Schimel KS (eds) Exchange of trace gases between terrestrial ecosystems and the atmosphere. Wiley, New York
26. Godfray HCJ, Coauthors (2010) Food security: the challenge of feeding 9 billion people. Greenhouse gas emissions from food loss and waste in the global food supply chain. Sci Total Environ 571:721–729. https://doi.org/10.1016/j.scitotenv.2016.07.041
27. Herrero et al (2016) Greenhouse gas mitigation potentials in the livestock sector. Nat Clim Chang 6:452–461. https://doi.org/10.1038/nclimate2925. http://www.nature.com/doifinder/10.1038/nclimate2925
28. Hristov AN, Coauthors (2013) Special topics—mitigation of methane and nitrous oxide emissions from animal operations: III. A review of animal management mitigation options. J Anim Sci 91:5095–5113. https://doi.org/10.2527/jas2013-6585
29. Hristov AN, Coauthors (2015) An inhibitor persistently decreased enteric methane emission fromdairy cows with no negative effect on milk production. Proc Natl Acad Sci 112:10663–10668
30. Hussain S, Peng S, Fahad S, Khaliq A, Hunag J, Ciu K, Nie L (2015) Rice management interventions to mitigate greenhouse gas emission: a review. Environ Sci Pollut Res 22:3342–3360. https://doi.org/10.1007/s11356-014-3760-4
31. Kumar Y, Berwal R, Pandey A, Sharma A, Sharma V (2017) Hydroponics meat: an envisaging boon for sustainable meat production through biotechnological approach: a review. Int J Vet Sci Anim Husb 2:34–39. https://www.veterinarypaper.com/archives/2017/2/1/A/1-3-25
32. Kuzyakov Y, Subbotina I, Chen H, Bogomolova I, Xu X (2009) Black carbondecomposition and incorporation into soil microbial biomass estimated by ^{14}C labeling. Soil Biol Biochem 4(1):210–219
33. Lu WF, Chen W, Duan BW, Guo WM, Lu Y, Lantin RS, Wassmann R, Neue HU (2000) Methane emission and mitigation options in irrigated rice fields in southeast China. Nutr Cycl Agroecosyst 58:65–74
34. Majumdar D, Kumar S, Pathak H, Jain MC, Kumar U (2000) Reducing nitrous oxide emission from an irrigated rice field of North India with nitrification inhibitors. Agr Ecosyst Environ 81(3):163–169
35. Malla G, Bhatia A, Pathak H, Prasad S, Jain N, Singh J (2005) Mitigating nitrous oxide and methane emissions from soil in rice–wheat system of the Indo-Gangetic Plain with nitrification and urease inhibitors. Chemosphere 58(2):141–147
36. Mohanty S, Nayak AK, Kumar A, Tripathi R, Shahid M, Bhattacharyya P, Raja R, Panda BB (2013) Carbon and nitrogen mineralization kinetics in soil of rice–rice system under long term application of chemical fertilizers and farmyard manure. Eur J Soil Biol 58:113–121
37. Mohanty S, Swain CK, Tripathi R, Sethi SK, Bhattacharyya P, Kumar A, Raja R, Shahid M, Panda BB, Lal B, Gautam P (2018) Nitrate leaching, nitrous oxide emission and N use efficiency of aerobic rice under different N application strategy. Arch Agron Soil Sci 64(4):465–479
38. Montagnini F, Nair PKR (2004) Carbon sequestration: an underexploited environmental benefit of agroforestry systems. Agrofor Syst 61–62:281–295. https://doi.org/10.1023/B:AGFO.0000029005.92691.79. http://link.springer.com/10.1023/B:AGFO.0000029005.92691.79
39. Naqvi SWA, Noronha RJ (1991) Nitrous oxide in the Arabian Sea. Deep Sea Res 38:871–890
40. Nelson ME, Hamm MW, Hu FB, Abrams SA, Griffin TS (2016) Alignment of healthy dietary patterns and environmental sustainability: a systematic review. Adv Nutr An Int Rev J 7:1005–1025. https://doi.org/10.3945/an.116.012567
41. Neogi S, Bhattacharyya P, Roy KS, Panda BB, Nayak AK, Rao KS, Manna MC (2014) Soil respiration, labile carbon pools, and enzyme activities as affected by tillage practices in a tropical rice–maize–cowpea cropping system. Environ Monit Assess 186(7):4223–4236
42. Nguyen Q, Bow J, Howe J, Bratkovich S, Groot H, Pepke E, Fernholz K (2017) Global

43. Pathak H, Aggarwal PK (2012) Low carbon technologies for agriculture: a study on rice and wheat systems in the Indo-Gangetic Plains. Indian Agricultural Research Institute, p xvii, 78
44. Paustian K, Lehmann J, Ogle S, Reay D, Robertson GP, Smith P (2016) Climate-smart soils. Nature 532: 49–57. https://doi.org/10.1038/nature17174. http://www.nature.com/doifinder/10.1038/nature17174
45. Pikaar I, Matassa S, Rabaey K, Laycock B, Boon N, Verstraete W (2018). The urgent need to re-engineer nitrogen-efficient food production for the planet. In: Managing water, soil and waste resources to achieve sustainable development goals, vol 7. Springer International Publishing, Cham, pp 35–69 http://link.springer.com/10.1007/978-3-319-75163-4_3.
46. Porter SD, Reay DS, Higgins P, Bomberg E (2016) A half-century of production-phase production of second generation biofuels: trends and influences, 16 pp. https://www.dovetailinc.org/report_pdfs/2017/dovetailbiofuels0117.pdf.
47. Reay DS, Davidson EA, Smith KA, Smith P, Melillo JM, Dentener F, Crutzen PJ (2012) Global agriculture and nitrous oxide emissions. Nat Clim Chang 2:410–416
48. Satpathy SN, Mishra S, Adhya TK, Ramakrishnan B, Rao VR, Sethunathan N (1998) Cultivar variation in methane efflux from tropical rice. Plant Soil 202(2):223–229. Science 327:812–818. https://doi.org/10.1126/science.1185383
49. Shepon A, Eshel G, Noor E, Milo R (2018) The opportunity cost of animal based diets exceeds all food losses. Proc Natl Acad Sci 115:3804–3809. https://www.pnas.org/content/115/15/3804.abstract
50. Smith et al (2013) How much land-based greenhouse gas mitigation can be achieved without compromising food security and environmental goals? Glob Chang Biol 19:2285–2302. https://doi.org/10.1111/gcb.12160
51. Smith P (2013) Delivering food security without increasing pressure on land. Glob Food Sec 2:18–23. https://doi.org/10.1016/j.gfs.2012.11.008
52. Smith P, Coauthors (2014) Agriculture, forestry and other land use (AFOLU). Climate change 2014: mitigation of climate change. In: Edenhofer O et al (eds) Contribution of working group III to the fifth assessment report of the intergovernmental panel on climate change, Cambridge University Press, Cambridge, United Kingdom and New York, NY, USA. https://www.ipccnggip.iges.or.jp/public/2006gl/pdf/4_Volume4/V4_04_Ch4_Forest_Land.pdf
53. Wassmann R, Papen H, Rennenberg H (1993) Methane emission from rice paddies and possible mitigation strategies. Chemosphere 26(1–4):201–217
54. Wilson NLW, Rickard BJ, Saputo R, Ho S-T (2017) Food waste: the role of date labels, package size, and product category. Food Qual Prefer 55:35–44

Chapter 10
Barriers to Adaptation of Climate-Smart Agriculture

10.1 Barriers to the Adaptation of CSA

The CSA is the apt response of the learned stakeholders of agriculture (researchers, farmers, political leaders, economists, environmentalists, etc.) to tackle the recent challenges of sustainable food security and climate change issues. The CSA is built on three pillars: productivity, resilience, and mitigation to climate change. Inventions, back-stepping, and interlinking of technologies are important for CSA like other approaches, but its adaptation is crucial for getting its long-term beneficial impact. The adaptation of CSA depends on the potentialities of technologies, its performance at the field level, and its acceptance by the farmers. Therefore, barriers to the adaptation of CSA could be technology-centric or user-performance centric or both. Another school of thought is that the adaptation barriers to CSA are primarily socio-economic both from "supply-side" and "demand-side." In this section, we will discuss the barriers to "land-based adaptation"; "policy-institutional framework"; "supply-side" (technology providers) and "demand-side" (technology user) chains; and from holistic "system-functioning" perspectives.

10.1.1 Barriers to "Land-Based Adaptation"

The "land-based" approaches mostly concentrate on mitigation to climate change and address either of the two pillars of CSA (productivity and resilience/mitigation). It primarily emphasizes growing of crops for biofuel, exploration of renewable sources of energy (sun, wind, water, etc.), and afforestation. In adaptation front, changing of cropping sequences/pattern, use of stress-tolerant cultivars with higher water and nutrient use efficiencies are recommended. However, many a time's adaptation-cum-mitigation practices/technologies are followed that includes the change in dietary pattern, use of solar and wind power at degraded wastelands, silvipasture in problem

© The Editor(s) (if applicable) and The Author(s), under exclusive license to Springer Nature Singapore Pte Ltd. 2020

P. Bhattacharyya et al., *Climate Smart Agriculture*, Green Energy and Technology,
https://doi.org/10.1007/978-981-15-9132-7_10

soils, etc. The competition for land for food security, production of crops for biofuel, providing ecosystem services, and maintaining biodiversity are the source of conflict and pose as barriers to CSA adaptation. The real barriers are trade-offs between land for food security and mitigation. The absence of clear cut water and land rights policy further aggravate the conflicts. Upscaling of the mitigation could be only possible through dedicated use of wastelands and or use of residues for biofuel production. It desperately needs regulation at the national level with transparent governance to resolve the conflicts. The participatory approach in sharing the land and water rights for food security and energy-crop production can be successful and could be implemented by certification using widely accepted standards. In the forest sector, regulation to secure land tenure and livelihood support to local people can overcome the barriers and could support the land-based mitigations. Regulation on "land-use" is more important in highly populated "resource-poor" countries. The barriers to "land-based" adaptation of CSA are inherently low in less populated "resource-rich" countries having less intensive agriculture. The scarcity of groundwater, lack of direct support to field infrastructure, institutional capacity, and livelihoods inhibit the adaptation. Till now, no systematic incentives are associated with land right for the production of biofuel crops, providing ecosystem services, preserving biodiversity, and carbon saving. Therefore, the farmers are in a dilemma to adapt CSA out rightly. But the opportunities exist shortly to incentives for following the CSA practices as people are more health-conscious; valued the environment; and urbanization is accelerating. Presently, the conservation of energy, water, biodiversity, and nutrients has been priced low to their environmental value. Further, the slow pace at which the land-use transformations from unproductive food crops to highly valued fiber, biofuel, and other commercial crops take place pose as barriers to CSA adaptation. Introduction of tax exemption and incentives by the government to use relatively unproductive land for mitigation purposes can reduce the barriers partially.

10.1.2 Barriers from "Policy-Institutional Frame-Work"

The institutional dimension of the CSA strategies started at the beginning of this decade (2011). It basically focused on the policy framework, knowledge development, and channelization of market information [8]. Gender equality and social aspects are also addressed in climate-smart agricultural practices. However, discrete agricultural services, asymmetric access to extension officers, and interrupted information flow became major barriers that inhibit the adaptation of CSA strategies. In most of the cases, the linkages of policy-institutions have been broken at the local level [36]. Knowledge structure and real-time market information very often failed at institutional dimensions, although market opportunities alone can make CSA technologies profitable and adaptable. Creating suitable market conditions would reduce the risk of the investor of CSA policies [2, 20]. We know that good technology is not alone sufficient for higher adaptation, it needs a profitable value-chain by using market opportunity [14].

As we know the developed countries give more emphasis on mitigation of GHG emissions rather than on adaptation of CSA. On the other hand, in developing countries, the majority of action for enhancing crop productivity is to ensure food security and resilience to climate change [4, 9, 17]. So, common incentives for adapting CSA would not work globally and would further generate conflicts and barriers for adaptation. Therefore, a methodological framework that systematically assesses the three pillars (productivity, resilience, mitigation) for CSA options is necessary for evaluation which is prerequisites to fix the incentives.

10.1.3 Barriers from "Supply-Side" and "Demand-Side" Chains

The "supply-side" refers to the inventors/producers of the technologies, whereas the "demand-side" indicates the end users of the technologies. These barriers are more or less common in developing as well as developed countries with region-specific variations of constituents of supply or demand chains. The demand (user) and supply (technology provider) side barriers and constraints to adaptation and diffusion of CSA are governed by the vulnerability of the people and socio-economic policy-planning at the country level.

A key barrier is the involvement of high-cost factors in technological innovations, especially at the early diffusion stage due to poor commercialization efforts by the public as well as private agencies. The investment to new innovative technologies of CSA as the production level itself is low and often suffers from uncertainty for future monetary gain [13, 27]. This is generally termed as "early-adopter-costs" that impacted both technology producers and users. The factors responsible for this higher cost of technology diffusion are relatively higher input prices, apparently, no subsidy available for adoption of new technology, and lack of willingness of users to pay higher premiums price to new products/services [24, 33, 38]. Insufficient skilled and trained manpower in industries at the initial stage of new product/services manufacturing is also added to the cost of new CSA innovations/practices [15, 28]. Moreover, the psychological barriers of the user (demand-side) to accept new CSA approaches also play a crucial role in large-scale adaptation [22]. The psychological acceptance of new/innovative technologies of consumers depends on the age and education level of the adopter [37]. Uncertainty and future risk factors, market fluctuations, the inertia of organizations, habitual routines, poor motivational leadership, and lack of foresight of leaders as well as institutions are the other key "demand-side" barriers of poor adaptation of CSA [7, 19, 35]. Conflicts of landowners and tenant-farmers over short-term gain of technology adoption is also a constraint for adaptation.

The typical barriers from "supply-side" (technology-produces) can be listed as financial barriers, avoidance of future risk (price volatility), and behavioral limitations. Sustainable development in contrast with short-term economic benefits to secondary-stakeholder poses barriers for adaptation. For example, the use of climatic stress-tolerant genetically modified crops/breeds (climate-proof technology) often raises controversies for adaptation in terms of specific commercial benefits or wider societal acceptance. Radical climate-proof technologies usually face greater inhibition for adaptation because of knowledge gap, lack of institutional and administrative capabilities, and difficulties for commercialization with existing resource-cum-knowledge base at the industry level [6, 21]. Market choice and competition with popular established product/service generally do not allow the adaptation of long-term CSA practices to small farm holding in developing countries. So, market forces indulge a key barrier to technology provider at the "supply-side" chain [12]. Moreover, input-subsidy to existing technology or product in most of the developing countries hardly encourage the acceptance of new (innovative) CSA practices having no or meager price support (subsidy for adoption) [38].

The barriers for adaptation of CSA are presented in a flow diagram for easy understanding (Fig. 10.1) showing the "supply-side" and "demand-side" chain. Four factors broadly identified in "supply-side" include lack of capital investment to technology/product development and commercialization; the relatively higher cost involved in product development; low immediate demonstration value as most of the CSA technologies/practices provide long-term environmental and economic benefit; and poor access to the potential consumers. On the other hand, eight (8) factors can be taken in the "demand-side" (technology users) chain. The factors are (i) low awareness of primary (farmers) and secondary consumers; (ii) low level of consumer demand at the initial stage of adoption; (iii) low short-term economic benefits for

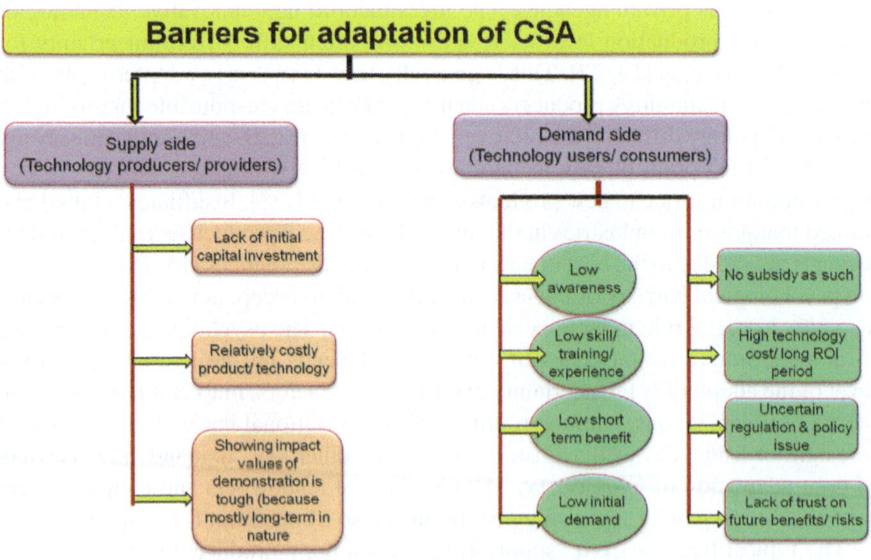

Fig. 10.1 Barriers to the adaptation of climate-smart agriculture

following the CSA practices; (iv) relatively higher cost of technology and long "pay-back" (ROI) period; (v) lack of skilled and trained manpower available for implementing the CSA practices precisely; (vi) no clear cut subsidy (liquid price support) as such for the adoption of CSA products/technology; (vii) lack of trust on the future benefits of CSA and associated risk; and (viii) uncertainly in regulation and policy issues related to CSA adoption (Fig. 10.1).

10.1.4 Barriers from "Holistic-System-Functioning"

The CSA is a pro-environmental technological approach. It needs genuine long-term policy support and regulations for higher adaptation. The common barriers to pro-environmental technology in general and CSA in particular, can be categorized into six (6) interlinked categories. The broader categories are economic, organizational, regulatory, behavioral, market-driven, and social (Fig. 10.2). The high initial cost of technology; low capital investment in technology development and large-scale commercialization; generally delay pay-back periods of pro-environmental technologies; uncertain returns; few hidden cost; competition in financing; low short-term benefit: cost ratio; and relatively higher implementation cost of CSA practices are economic barriers (constraints) to the adaptation of CSA. Regulation and institutional barriers are directly related to economic constraints. Low institutional support makes innovative CSA products/technologies/services costly. In true sense, until now there is low-regulation framework for CSA in most of the countries that slower the adaptation. The top-down approaches without considering the farmers' knowledge and experiences in research and development issues and subsequent diffusion strategy of CSA inhibit the adaptation. Sometimes it is also difficult to formulate the regulation framework of CSA and execute it at local as well as national level due to non-availability of perceived standard for assessment of CSA practices. Further,

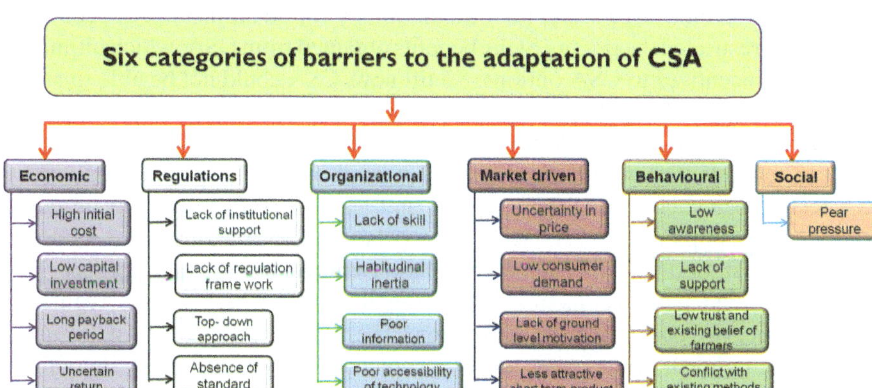

Fig. 10.2 Categories of barriers to the adaptation of CSA

the use of excessive scientific jargon to the new radical/innovative CSA technologies poses some limitations for general acceptance by farming communities. The organizational barriers include lack of trained manpower; skilled persons/technicians (both at industry and farm level); poor information flow from center to local implementing agencies; and inherent organizational-inertia hinder the adaptation. The organizational barriers are interlinked with economic as well as regulatory constraints. The fourth and other important barriers come from the consumers (technology users) which are market-driven. Uncertainty in output-price; generally low consumer demand as CSA has driven benefits that are usually gained at mid- and long-term time frame (low attractive product/technology at short-term basis); and lack of ground-level motivation for the adoption of CSA are the barriers for adaptations. Consumer/market-driven barriers are although linked with organizational constraints but more influenced by behavior of farmers (consumers) and social factors (Fig. 10.2). The last two interlinked and overlapping categories of barriers are social and behavioral constraints. Both the categories are user/consumer mediated, i.e., in the "demand-side" chain. Low awareness; lack of immediate financial support (subsidy/incentives); the low trust of CSA practices (which are generally less practically validated in various locations); the traditional belief of farmers (contradicts with CSA as such); and conflicting with the existing method of farming-practices hinder the adaptation of CSA (Fig. 10.2) [8, 10, 27, 28].

10.2 Causes of Barriers to the Adaptation of CSA

The nature of CSA and its long-term impact on the environment and economy are the causes of barriers. Any investment needs an assured profit in short- as well as the long-term time scale. But generally, investors are reluctant to support CSA adaptation because they are not sure about return and they have to wait a long payback period. Therefore, the nature of the CSA approach itself poses as the barriers to its adaptation on a large scale. Poor methodology for quantification of ecosystem services has restricted the pricing of the benefits obtained from CSA, which ultimately inhibits the incentives to CSA options. So till now, CSA could not be able to attract international and national financing. Lack of incentives to CSA practices and low financial support are the key barriers to its adaptation.

Given less attention to the development of rural infrastructure to accommodate CSA technologies, contextual-perspective, and less effort to bridge private-public partnerships to implement CSA causes barriers both in organizational and economic categories. Institutional dimensions and market-knowledge of CSA options' have not been studied and documented widely in literature which causes regulatory and consumer-driven barriers to adaptation. The knowledge gaps exist in understanding the political preferences, social acceptance, local needs, cultural conflicts, and historical-legacies in uptake/adoption of CSA practices. These gaps generate some inherent barriers to adaptation in behavioral and social categories. The lack of documentation on site-specific contextual limitations and interactions of private-public sectors during scaling up of CSA technologies invites barriers. The relevant data on farmers' field validation of CSA practices as a whole are lacking globally which makes the technology less trustworthy to farmers [25, 26].

Further, due to the lack of immediate short-term benefits of CSA technologies, the private sector often thought that the investment in this sector is a negative business proposition. However, transparent public-private partnerships may create attractive long-term business opportunities for up-scaling CSA options. Lack of evidence and documentation of successful private model for CSA (like, Manobi-private company: in promotion of CSA in vulnerable communities) promotion hinders private-public cooperation to support CSA which is the biggest financial barrier [11, 29, 31, 32]. The understanding of type and kind of cooperation between public and private institutions is required for adaptation of CSA strategies that are still limited. Poor investment to rural roads, cold storage-facilities, easy transport to market, and irrigation network is the general infrastructural issues in agriculture but those, in turn, cause barriers to CSA adaptation by increasing the vulnerability of the people of the region. For example, Rinaldi et al. [34] showed that irrigation supported the adaptation to crop-diversification and up-scaling of CSA practices. Last but not the least, although theoretically, the approaches of CSA are excellent, but in real sense there is still a lack of policy framework for integrating food security in the challenging climate change scenarios.

10.3 Possible Options to Overcome the Barriers to the Adaptation of CSA

The climate action for food security begins after the Paris Agreement both by "resource-poor" and "resource-rich" countries through various initiatives which include policies and frameworks for CSA adaptation. The Nationally Determined Contributions (NDCs) by various countries are the major commitment toward that. It is an indication of the efforts by global communities toward climate change mitigation and domestic adaptation of climate change policies. The NDCs of south Asian countries show their commitments to mitigate climate change and restrict the global temperature to rise up to 2 °C through adapting climate-proof practices and or CSA approaches. Meta-analysis showed that mitigation technologies that facilitate income-generation to farmers and sustain food security are basically adapted by the communities in South Asian regions. The lesson learned from the experiences that the climate-smart value-chain which is self-sustaining and of higher efficiencies have been adapted widely. This message should also be carried forward during the CSA adaptation endeavor. The ecosystem-based landscape model facilitated the adaptation of climate-resilient technologies in Asian countries in the last 10–15 years. The cost of adaptation must be incorporated in CSA policies which is a take-home message derived from the performance analysis of NDCs in south Asian countries. The countries also used a different source of financing for the execution of climate change mitigation policies that include an international, national, private, public source of fund. This model of adaptation-financing also could be used in CSA diffusion.

In this connection, it is to be understood that the CSA which enhances productivity and requires relatively low cost for initial implementation is easily acceptable in developing countries. Therefore, the nature of CSA technology per se is important for its early adaptation. For example, in-situ algal production in rice fields and residue management through farm mechanization (happy seeder, zero-tilled seed drill, etc.) in intensive rice-wheat cropping system are adapted by farmers. Direct payment for carbon-saving, immediate incentives for practicing CSA may reduce the economic barriers (i.e., the transaction costs in small farm holding) of the majority of Asian farmers that could facilitate the adaptation of CSA practices [1].

There are some inherent barriers to developing countries to adopt CSA. Specifically, those are having small farmers (majority farm holding is small and farmers are either small or marginal; <2 ha land) and long market-value-chain (a lot of middlemen and weak value-chain). So, structural interventions to modify farm size (to optimum scale) through co-operative and or contact-farming may trigger the autonomous adaptations of CSA by farmers and could enhance their adaptive capacity [16].

Eliminating gender issues, greater empowerment of women, strong market value chains are few effective options to reduce social and behavioral barriers of adaptation to CSA. Rural employment generation via climate change mitigation activities and through "labor-compensation" model could be effective to arrest social and behavioral barriers of adaptation to CSA in "resource-poor" countries.

There are options available to overcome the economic, social, organizational, behavioral, and regulatory barriers to the adaptation of CSA (Table 10.1). The investment in technology development and commercialization can effectively reduce economic barriers. Strengthening the supply value-chain of CSA products and start-up support to young entrepreneurs of CSA technologies would be beneficial. The "user-centered' approach both at product development and regulation is needed to make CSA more user friendly. The bottom-top approach for on-field demonstration of CSA practices with direct incentives is effective. The climate-awareness, training for handling CSA equipment, skill development at rural scale, and encouraging rural youth for adopting CSA practices are a real investment to CSA. However, there is a debate about the cost and effectiveness of CSA technologies but we believed that policy, finance, and consumer preference are the key barriers to the adaptation of CSA practices. In the Table 10.1, we elaborated the options to overcome the barriers in five different categories (economic, social, regulatory, consumer, organizational) for easy understanding; however, all the barriers and the options to overcome them are interlinked and complementary with each other. Policy framework, regulation, technology development, and consumer awareness have to be addressed simultaneously for higher adaptation of CSA (Table 10.1).

Table 10.1 Options to overcome the barriers to the adaptation of CSA

Sl. no.	Barriers	Options/strategies
1	Economic	A better balance of benefit: cost ratios of technologies through the supply chain
		Long-term investment by international funding agencies
		"Start-up" support for technology innovation and commercialization
		"User-Centered" creation/innovation of CSA products/technology/services
		Tax-breaks to research and development for CSA technology generation
2	Regulatory	"Bottom-top" approach for technology diffusion
		"User-Centered" policy and regulation
		Development of CSA business-module
3	Organizational	CSA technology validation at farmers' field
		Demonstration for showing the short-term benefit of CSA
		Incorporation of technology development and adaptation cost in policy framework itself
4	Market-driven/consumer	Greater access to consumer
		Consumer-Segment based promotion of specific CSA options
		Evidence-based incentives for following the CSA practices
		Priced the carbon-saving technologies and ecosystem services in a systematic well-accepted way
5	Behavioral/social	Training of rural youth; and capacity building
		On-farm demonstration of CSA practice in a farmers' participatory mode
		Climate-awareness campaign; CSA-Education; and direct incentives in monetary term for following the CSA practices at the local level

Sources Edler [18], OECD [30], Wubben et al. [39]

10.4 Trade-Off Between Food Security and Adaptation of CSA

i. There are potential short-term adverse effects on food, water, and nutritional security in response to the land-based adaptation of CSA. The risk of adverse side-effects may increase with scaling up of cropland for bioenergy production and higher deployment of fertile land to capture and storage of carbon.

ii. However, the "sustainable development pathways" (particularly in SSP1) mentioned regulation of proper land-use to reduce environmental trade-offs. The regulation should aim to break the vicious cycle of "poverty-land degradation-GHG emissions" comprehensively on a system basis. For example, lower the non-vegetarian diet (meat), change in growth and food-consumption pattern, and encourage the local markets. The international trade also has the potential to regulate food prices as well as to reduce the GHG emission and vulnerabilities of farmers to climate change vagaries (floods, drought, cyclones, etc.).
iii. The burden of carbon-emission reduction should be logically distributed to responsible sectors (energy, industry, agriculture, etc.). Overburdening the agriculture sector (land sector) could adversely affect food security and other ecosystem services in the near future. The risk of mitigation failure from the land sector also should be accounted for as the agriculture primarily dependent on climatic factors which are also supposed to be changed.
iv. Trade-offs exist in climate mitigation with biodiversity. The food and land-security issues are rarely spelled out into CSA or other climate policies. For example, the downstream river ecology and fish production could be impacted negatively by the installation of hydropower stations upstream. Intensive use of rivers for transportation may hamper fish productivity. The endangered species and biodiversity are often disrupted with the installation of large-scale wind-turbine and or solar-farms.
v. The carbon market and incentives for carbon saving are more cost-effective than taxation or punishment for higher emission. However, it is relatively less validated and tested in agriculture and food sector globally. So, the trade-off remains a debatable issue unless the agricultural GHG emissions mitigation is transparently included in mainstream climate policy. In this context, there is a scope for ensuring the equity in carbon pricing by mixing both market and non-market mechanisms and proofing the leakage of multi-sector emissions.
vi. Lack of investment in assessment, monitoring, evaluation of impacts analysis of different CSA policies across different sectors generates "trade-offs" and "knowledge gap." So, it weakens the understanding of the effectiveness of the policy at agriculture, forestry, bioenergy, and related land-use management levels. The majority of the climate-measures and land-use changes are long term in nature and need to be evaluated in the socio-economic frame. Therefore, it is difficult to compare with energy and industry sectors as those are primarily technology-driven.
vii. The poor linkage among institutional context, CSA technology innovation framework, and socio-economic acceptance causes the trade-off between food security and mitigation to climate change. These conflicting thoughts (preferences of adaptation-mitigation vs. food security) and apparent delay of the pay-back period on investment on CSA strategies cause the barriers to adaptation at the local level.

viii. Above all, many researchers and policymakers think that there are trade-offs that exist in the conceptual levels of CSA, i.e., among the three pillars (productivity, resilience, and mitigation) of the CSA approach itself. In order to scaling up the CSA, the political and institutional dimensions of CSA have to be looked into the context of historical, cultural, and economic perspectives.

10.5 Summary and Messages

- Climate change adaptation is necessary for sustainable food, agricultural, and environmental security. The CSA has been playing a key role in this juncture. In general, for wider adaptation of any climate policy or CSA needs proper integration of adoption of stress-tolerant cultivars; practicing sustainable soil-water-nutrient management; development of skill, knowledge, and climate-literacy; and building of transparent policy and institutional setups [23].
- Technological innovation (climate stress-tolerant cultivars/breeds, etc.), climate-risk responsive institutions and climate policies, crop insurance, incentives to ES (ecosystem services), participatory community-based strategies to tackle sudden climatic-hazards are back-bones of CSA adaptation.
- Different countries already have their national policies to address the climate change issues; however, there is a need for greater coherence and coordination among various levels of institutions (international and national) for implementing CSA globally.
- Financing to CSA adaptation in particular and climate-action, in general, is still low throughout the world, whether it is international agencies, national institutions, and private organizations that need to be strengthened.
- Standard pricing of ecosystem services, carbon saving due to adapting CSA is a prerequisite for its adaptation. Transparent and widely accepted methodologies must be developed and validated across the regions of the world [3, 5]. Impact analysis of CSA both at mid- and long-term time frame is important. Not only extension research but "user-centric" innovative technology-development researches are important for adaptation and scaling up of CSA.
- System reliance in agriculture is necessary by integrating global challenges of food security and climate change. In this context, the total system transformations from "provider-centered" to "user-centered" are needed.
- Effective farmers' participation in the implementation of CSA technologies in compliance with "Government climate policy" is the key to the adaptation of CSA.
- In nutshell, the adaptation of CSA should be tackled at system level, to provide justice to complex issues like hunger, livelihoods, poverty, malnutrition, economy, sustainability, and environment under climate change vagaries.
- It is our collective responsibility to mitigate climate change as we are "donor" as well as the "receptor" of climate change-related risk. Only vulnerability differs based on our present coping capacity, but we should remember that the "coping

capacity" to climate change is highly fragile and purely a relative term. Therefore, it needs an inclusive policy, considering favorable to vulnerable people/regions, by providing equal opportunity to the small holders, women, and poor communities in the world.

References

1. Aggarwal PK, Jarvis A, Campbell BM, Zougmoré RB, Khatri-Chhetri A, Vermeulen S et al (2018) The climate-smart village approach: framework of an integrative strategy for scaling up adaptation options in agriculture. Ecol Soc 23(1):14
2. Ajayi OC, Catacutan D (2012) Role of externality in the adoption of smallholder agroforestry: case studies from Southern Africa and Southeast Asia. In: Sunderasan S (ed) Externality: economics, management and outcomes. Nova Science Publishers Inc., New York, NY, USA, pp 167–188
3. Amjath-Babu T, Aggarwal PK, Vermeulen S (2018) Climate action for food security in South Asia? Analyzing the role of agriculture in nationally determined contributions to the Paris agreement. Clim Policy 19:1–16
4. Anand A, Khetarpal S (2015) Impact of climate change on agricultural productivity. In: Plant biology and biotechnology: plant diversity, organization, function and improvement, vol 1. Springer, New Delhi, India, pp 729–755
5. Bandara JS, Cai Y (2014) The impact of climate change on food crop productivity, food prices and food security in South Asia. Econ Anal Policy 44:451–465
6. Bessant J, Öberg C, Trifilova A (2014) Framing problems in radical innovation. Ind Mark Manag 43(8):1284–1292
7. Blok V, Long TB, Gaziulusoy AI, Ciliz N, Lozano R, Huisingh D, Boks C et al (2015) From best practices to bridges for a more sustainable future: advances and challenges in the transition to global sustainable production and consumption: introduction to the ERSCP stream of the special volume. J Clean Prod 108:19–30
8. Bogdanski A (2012) Integrated food–energy systems for climate-smart agriculture. Agric Food Secur 1(1):9
9. Branca G, Lipper L, Sorrentino A (2015) Cost-effectiveness of climate-related agricultural investments in developing countries: a case study. New Medit 14:4–12
10. Brunke JC, Johansson M, Thollander P (2014) Empirical investigation of barriers and drivers to the adoption of energy conservation measures, energy management practices and energy services in the Swedish iron and steel industry. J Clean Prod 84:509–525
11. Chandra A, McNamara KE, Dargusch P, Damen B, Rioux J, Dallinger J, Bacudo I (2016) Resolving the UNFCCC divide on climate-smart agriculture. Carbon Manag 7(5–6):295–299
12. Costa-Campi MT, Duch-Brown N, Garcia-Quevedo J (2014) R&D drivers and obstacles to innovation in the energy industry. Energy Econ 46:20–30
13. Cullen R, Forbes SL, Grout R (2013) Non-adoption of environmental innovations in wine growing. N Z J Crop Hortic Sci 41(1):41–48
14. Dalohoun DN, Hall A, Mele PV (2009) Entrepreneurship as driver of a 'self-organizing system of innovation': the case of Nerica in Benin. Int J Technol Manag Sustain Dev 8:87–101
15. del Río González P (2005) Analysing the factors influencing clean technology adoption: a study of the Spanish pulp and paper industry. Bus Strateg Environ 14(1):20–37
16. Dev SM (2012) Small farmers in India: challenges and opportunities. Indira Gandhi Institute of Development Research, pp 8–14
17. Dougill AJ, Whitfield S, Stringer LC, Vincent K, Wood BT, Chinseu EL, Steward P, Mkwambisi DD (2017) Mainstreaming conservation agriculture in Malawi: knowledge gaps and institutional barriers. J Environ Manag 195:25–34

18. Edler J (2007) Demand-based Innovation Policy Manchester Business School Working Paper, No. 529, Manchester, UK
19. Faber A, Hoppe T (2013) Co-constructing a sustainable built environment in the Netherlands—dynamics and opportunities in an environmental sectoral innovation system. Energy Policy 52:628–638
20. Fischer HW, Reddy NLN, Rao MLS (2016) Can more drought resistant crops promote more climate secure agriculture? Prospects and challenges of millet cultivation in Ananthapur, Andhra Pradesh. World Dev Perspect 2:5–10
21. Hall J, Vredenburg H (2003) The challenge of innovating for sustainable development. MIT Sloan Manag Rev 45(1):61
22. Hoffman AJ, Henn R (2008) Overcoming the social and psychological barriers to green building. Organ Environ 21(4):390–419
23. Hurlbert MA, Gupta J (2019) An institutional analysis method for identifying policy instruments facilitating the adaptive governance of drought. Environ Sci Policy 93:221–231
24. Kemp R, Volpi M (2008) The diffusion of clean technologies: a review with suggestions for future diffusion analysis. J Clean Prod 16(1):S14–S21
25. Kijima Y, Otsuka K, Sserunkuuma D (2011) An inquiry into constraints on a green revolution in Sub-Saharan Africa: the case of NERICA rice in Uganda. World Dev 39(1):77–86
26. Lipton M, Ahmed I (1997) Impact of structural adjustment on sustainable rural livelihoods: a review of the literature
27. Luthra S, Kumar S, Kharb R, Ansari MF, Shimmi SL (2014) Adoption of smart grid technologies: an analysis of interactions among barriers. Renew Sustain Energy Rev 33:554–565
28. Montalvo C (2008) General wisdom concerning the factors affecting the adoption of cleaner technologies: a survey 1990–2007. J Clean Prod 16(1):S7–S13
29. Msalya G, Lie H, Mfinanga V, Ringheim AS, Sandvik R, Åsli M, et al (2016) Public-private partnership for sustainable production and marketing of goat's milk in light of climate change. In: Climate change and multi-dimensional sustainability in African agriculture. Springer, Cham, pp 505–524
30. OECD (2014) Stimulation demand for innovation. OECD, Paris, France
31. Pulkkinen H, Roininen T, Katajajuuri JM, Järvinen M (2016) Development of a climate choice meal concept for restaurants based on carbon footprinting. Int J Life Cycle Assess 21(5):621–630
32. Quail S, Onyango L, Recha J, Kinyangi J (2016) Private sector actions to enable climate-smart agriculture in small-scale farming in Tanzania. In: Climate change and multi-dimensional sustainability in African agriculture. Springer, Cham, pp 525–551
33. Reinstaller A (2008) The technological transition to chlorine free pulp bleaching technologies: lessons for transition policies. J Clean Prod 16(1):S133–S147
34. Rinaldi SM, Peerenboom JP, Kelly TK (2001) Identifying, understanding, and analyzing critical infrastructure interdependencies. IEEE Control Syst Mag 21(6):11–25
35. Roelofsen M, Blok V, Wubben EF (2015) Maintaining the CSR-identity of sustainable entrepreneurial firms. In: Corporate social responsibility and governance. Springer, Cham, pp 63–88
36. Shames S, Heiner K, Kapukha M, Kiguli L, Masiga M, Kalunda PN, Ssempala A, Recha J, Wekesa A (2016) Building local institutional capacity to implement agricultural carbon projects: participatory action research with VI agroforestry in Kenya and ecotrust in Uganda. Agric Food Secur 5:13
37. Tey YS, Brindal M (2012) Factors influencing the adoption of precision agricultural technologies: a review for policy implications. Precis Agric 13(6):713–730
38. Weiss C, Bonvillian WB (2013) Legacy sectors: barriers to global innovation in agriculture and energy. Technol Anal Strateg Manag 25(10):1189–1208
39. Wubben E, Runge N, Blok V (2012) From waste to profit: an interorganisational perspective on drivers for biomass valorisation. J Chain Netw Sci 12(3):261–272

Chapter 11
Mainstreaming of Climate-Smart Agriculture

11.1 Policies

11.1.1 Concept of CSA Policies

The concept of "Green Economy" is a primary component of "Sustainable Development" evolved since 2012 in the Rio summit of the United Nation's Conference. The call-in that summit was primarily for the eradication of hunger through sustainable development of crops, grassland, forestry, livestock, aquaculture, and forestry which should be economically viable. Another point must be noticed here, that when we are talking about "sustainable agriculture", the conservation of soil, water, plant and animal genetic resources, biodiversity, and ecosystem also should be taken into consideration. The copping mechanism to natural disasters and the enhancement of resilience to climate change are an integral part of that approach [20]. The three pillars of CSA, i.e., sustainable productivity, increasing resilience, and reducing GHGs emissions are also by and large matching with the "Green Economy" concept. Therefore, we can tell that the CSA and "Green Economy" are complementary concepts. So, in a boarder sense, greening the economy with climate-smart agriculture approaches could be a realistic pathway to achieve the sustainable development goal [7].

Agriculture is the biggest sector that utilizes and conserve natural resources. Globally, agriculture uses 70 and 40% of water and land resources, respectively, and accounts for around 30% of global GHGs emissions. Therefore, agriculture is the key to climate-smart development and green growth globally. It is also directly associated with food and livelihood security and human welfare. The CSA should consider the sustainable growth of food, pasture, fiber, fuel, and timber economically that address the poverty reduction in one hand and enhancing the resilience of the system to climatic hazards. It helps in maintaining ecosystem services, protecting natural resources with the objective of green economic growth. Climate change mitigation is the other important aspect that must be looked into for framing policies

© The Editor(s) (if applicable) and The Author(s), under exclusive license to Springer Nature Singapore Pte Ltd. 2020

P. Bhattacharyya et al., *Climate Smart Agriculture*, Green Energy and Technology,
https://doi.org/10.1007/978-981-15-9132-7_11

for CSA. In a nutshell, CSA as a productive, resilient, resource-efficient, and low carbon emitting approach needs a special policy-framework to implement in local and regional level.

11.1.2 Previous Policies of CSA

The CSA concept was first introduced in the Hague conference of the UN, during 2010 on the issue of "Agriculture-Food Security-Climate Change". Soon after that in 2011, the FAO launched the EPIC (Economic and Policy Innovations for Climate-Smart Agriculture (EPIC)) program. In that program, Malawi, Vietnam, and Zambia worked with European counterparts for developing and implementing the CSA practices with five major themes. Those were,

(i) Identification and ranking of feasible agricultural technologies and prioritization of site-specific viable CSA portfolios based on proven food security, adaptation, and mitigation benefits.
(ii) Assessment of the barriers to adopting CSA practices at the local level and finding policy and institutional options to overcome the difficulties.
(iii) Risk assessment and impact analysis of climate change on investment, financing, and policy-framework.
(iv) Capacity building at the country level by making larger coherence between human resources development and agricultural technologies dissemination.
(v) Establishing greater linkage between international funding agencies with partner countries for higher investment for scaling up CSA activities.

The success of the previous climate change-related program based on the very basic principle that those climate-resilient strategies had been interlinked with the then-existing rural development and agricultural policies [13]. The same principle was followed in PPCR (Pilot Program for Climate Resilience) which was directly supported by CIF (Climate Investment Funds). The "Climate Investment Fund" straightway approved a one billion US dollar grant for mainstreaming climate-resilient strategies into the core program. In which, water and agriculture management features were common to most of the participating countries in PPCR. And the program was reasonably successful. Those teach us the lesson that the PPCR-model could be used to implement the CSA in the future. The functioning and implementation strategies of the "Green Climate Fund" approach are also useful in this context which can show the effective pathways for implementation of CSA at the country level.

11.1.3 Principles and Elements of CSA Policies

The policies of CSA should integrate food security, agricultural development, and climate change issues coherently by putting equal importance. The CSA approaches have to be mainstreamed with the core governmental policies at the national level. Public, as well as private support, must be ensured by putting transparent benefit-sharing conditions in the policy-framework itself. The policies must be coherent with national ongoing policies in the long-run and never-conflict with the existing policies in the short run. The policies must be coherent with the existing long-term policies like UNFCC (United Nations Framework Convention on Climate Change), National Communications; NAPAs (National Adaptation Programmes of Action), and NAMAs (National Appropriate Mitigation Actions) [6]. Unlike other developmental policies, the CSA policy-framework also should take care of gender issues, protections to climate-vulnerable poor populations, and ecosystem-sustainability. The policies must ensure the sharing of the credits/benefits (in monetary terms) of CSA both to the weaker and unprivileged (women in particular) sections of the society as they are going to be worst sufferers due to climate change vagaries.

The CSA policies must be planned and implemented at the national and sub-national levels, but the national policies very often influenced by international agreement and support mechanisms. Thus, the national CSA policies are impacted directly by the global agreement on climate change mitigation and adaptation strategies. After the Rio summit and onwards, agriculture has been getting importance on climate change policies by the UN (United Nations) and other climate change dealing organizations in the world. Despite that also agriculture is still a neglected sector in terms of global-climate-financing [10]. However, in recent past, many developing countries particularly in south-east Asia and Africa have shown interest in agricultural mitigation and adaptation action to climate change. The developing countries also took interest in mainstreaming the CSA policies to their respective sub-national and nationally owned strategies and policies. They have taken those issues with respect to food security and climate change perspective and have been tried to incorporate those into their domestic policies by using international mechanisms/instruments to meet the challenges [10, 14]. But the basic problem faced by the developing countries is the conflict between the "removing public subsidies" and providing "incentives for adopting CSA practices" in their domestic policies. In general, most of the "public subsidies" are to ensure food security by increasing production and distribution with a relatively short-term goal. On the other hand, "incentives to adopting CSA practices" deal with the environment and ecological issues on a long-term basis which is not apparently lucrative for low- and middle-income groups in developing countries. Therefore, reallocating the resources from "good/service-subsidies" to "incentives for adopting CSA practices" is difficult and has been facing a lot of political and social hindrance. However, there are few effective tools to empower the rural mass and build-up the capacity of farmers through real-time input-output pricing; rural credit initiatives; support to rural youth for start-up and investment for public interest; door-step extension services; safety-net programs [10]; and providing equal property right to men and women. These initiatives could bring the changes in approaches and modify the production system which ultimately is a part of CSA policy.

In this connection, one of the principles of FAO with respect to climate change agriculture is the approach of "No-Regrets", which is not only sustainable but also easily acceptable by both the developed and developing countries. The approach specifically emphasized the sustainable improvement of agriculture, livestock, forestry, and fisheries even in the absence of climate change vagaries, so that the adoption of CSA practices by the farmers and other stakeholders would be relatively easier. At the same time, we should also remember the site-specific recommendation of CSA practices that should address both climate change mitigation and adaptation together. For example, zero tillage practices along with the addition of crop residues as organic matter to soil could increase the adaptive capacity of farmers in moisture stress conditions and also help in climate change mitigation by storing more carbon in the soil. Similarly, site-specific small-scale integrated farming system (putting together crop and animal components with the recycling of organics in the system itself) with rainwater harvesting is an excellent CSA practice for small and medium farmers in south-east Asia that has both adaptive and mitigation capacity to climate change. Therefore, integrated approaches should be taken into agricultural-policy-framework that would sustain long-term productivity and reduces the trade-offs and at the same time enhances synergies between adaptation and mitigation to climate change. In this context, we have to remember that the FAO's "No-Regret" principal only would be successful in CSA policies if we put the major emphasis on sustainable productivity enhancement, which is the first pillar of the CSA approach. Therefore, policies for intensifying sustainable production are vital for the effective implementation of CSA. And, production-intensification policies should be in synergies with watershed management and landscape restoration on a sub-national or state or national level. Technological innovation and dissemination on increasing efficiencies of water, soil, nutrient, and energy must be embedded in the policies. Other important aspects are infrastructural development for storage, market-information, reducing food loss and waste, risk management, and increasing the efficiencies of the food-value chain. Last but not the least, the weather forecasting; real-time agro-advisory and crop-insurance are the keys to CSA policies that must be matched with a national, political, and social framework.

The planning for expenditure is equally important as technological back-stepping in CSA policy-framework. Coordination and transparency among local and national level agencies are important for CSA expenditure-planning. Regulatory framework and legal perspective are another important factor for CSA planning as the approaches include the environment, ecology, and food security all together. Therefore, proper understanding and sharing of responsibilities among the agencies of cross-sector are the prerequisites of CSA. Proper linkages among the "land-use regulation"; "disaster risk management"; and "social-safety-networks" at the country level could be useful for mainstreaming the CSA strategies at national policy. Provisions must be created for allocating the budget of CSA policies in a national-annual plan. And responsibilities should be fixed for public and as well as private line departments through the transparent legal and regulatory mechanism. It has to be a part of the national core agricultural policy. So, the public money has to be used with proper alignment of line departments in the respective financial allotment limit. Investment and financial allocation also should consider the charges of maintenance of CSA practices on a long-term sustainable basis. As environment-protection is an international issue but must be tackled at the local/country level, hence international and nationally integrated investment-framework could be established for financing CSA activities. In this connection, different public, as well as private sources, should be utilized for raising funds. Financial flow from the top (i.e., center) to down, (i.e., local private/public administration) should be timely and transparent at the national level. Decentralization is the key to the success of CSA policy implementation. As many times, public money needs to be used, so periodic review of financial utilization and technological intervention is important for CSA. Review, regulation, and support system should be described in policies itself and implemented meticulously for the successful implementation of CSA policies.

Before going to the subsidy issue and trade-offs between "public subsidies" and "incentive to follow the CSA practices" at the national level, we should clearly understand the three key elements of CSA policies. Those three elements are (i) protection of "land-tenure rights," (ii) building confidence in "private sector investment," and (iii) making the stakeholder well aware of "realizing the potential of marketing of CSA products/technologies" both at national and international level [10]. Securing land tenures right to farmers or other stakeholders by investing in land and water management on a sustainable basis is crucial. It facilitates long-term investment in CSA practices. A professional business environment must be created by the government at the national level through a simple and transparent tax system, financial regulation, and easy lending-conditions, so that private investors have the confidence to invest in CSA planning. Last but not the least, good transportation and market-information to farmers, regulating-market facilities for CSA products/technologies, and healthy competition among the agro-enterprises are vital for the successful implementation of CSA planning-framework.

11.1.4 Trade-Off Between "Subsidies" and "Incentives to CSA Practices"

Agricultural produce is mostly perishable and often having a volatile market price. The production of farms primarily depends on climatic conditions which are changing at an alarming rate. Moreover, the majority of the farming community all over the world are poor and vulnerable to climate change and market fluctuations. Therefore, most governments of developing or even economically developed countries are concerned about support prices to farm produce to protect the resource-poor farmers and ensuring food security of the country. The governments generally procure a certain quantity of farm produce from the farmers through some agencies at a predefined guaranteed price called "support-price". It is a kind of subsidy provided by the government to farmers on the marketable produce. This certainly helps the vulnerable community; a kind of assured support to the needy in the climate change scenario.

There are other types of subsidies existing in many countries in terms of incentives to use of technologies, implements, fertilizers, water, energy, seeds, etc. that could be more effective in case of climate change and market failures. Supports also provides to use of social-capital and social-marketing. Input-subsidies are still effective, widely used, and acceptable in "resource-poor" countries. However, often there are conflicts between "input-subsidy" and "support-price" to the produce. In this context, a part of "price-support" and or energy/fertilizer/water-subsidies could be effectively redirected toward CSA policies by providing "incentives to CSA practices". These can be done by putting values of environment-protection generated by the CSA practices. For example, subsidies may be given on the timely/efficient use of water and fertilizer instead of using the total quantity of fertilizer/water used. Regulation and efficient use of fertilizer and water have a crucial role in adapting and mitigating climate change and could be a part of the policy for implementing CSA [10]. More so, the limit of fertilizer use per hectare could be mentioned in the policy up to which the subsidies would be given so that overuse or exploitation of inputs could be reduced (e.g., in Tanzania, fertilizer subsidy is allowed up to 50 kg per hectare use only). In this context, we also should consider that a certain amount of fertilizer use is necessary for maintaining crop productivity and soil carbon storage; in which fertilizer subsidy has been playing an important role in countries with "low-input-output" agricultural systems. Soil carbon sequestration increases resilience and mitigates climate change. Advisory services for timely application of fertilizers and water are equally important as a subsidy to those resources for effective utilization and plays an important role in CSA. However, real-time weather forecasting, extension services, and access to information and resources are to be ensured for effective resource utilization. It could bridge the trade-off between "subsidy" and "incentive to CSA practices". In contrast, subsidy to the total amount of fertilizer use in the countries where the fertilizer application rate is already high should not consider as the climate-smart approach. The obvious reason is that subsidy to fertilizer or electricity for pumping groundwater irrationally in those regions where already intensive agriculture has been practicing lead to an overuse or misuse of fertilizers. It would further degrade soil and environment and at the same time deteriorate the long-term crop productivity. In intensive agriculture zones in a country or in a specific region "incentives to practicing CSA" should be encouraged. For example, farmers may be given rewards for providing ecosystem services and carbon storage either directly or indirectly. Providing subsidies to the use of efficient resources and machinery (e.g., happy seeder, zero-tilled seed drill, etc.), stress-tolerant cultivars (drought, submergence of salinity tolerant varieties) and technologies (e.g., direct-seeded rice in water scares region, rainwater harvesting, participatory seed production at the village level, etc.) are indirect support.

So, now it is clear that "subsidy" to input for ensuring food production and food security is the prime concern to economically poor countries, whereas, the economically rich country could give more emphasis to "incentives for adopting CSA practices". But even in a developing country, subsidy to fertilizer and electricity for pumping of groundwater for irrigation for the long-run lead to an imbalance in soil fertility and groundwater depletion as evident from north-west India, Pakistan, and Uzbekistan in last 2–3 decades. However, removing subsidy to basic inputs of agriculture in those countries is a socio-political issue, where environmental issues are hardly considered. But fortunately, in the last decades or so due to UN initiatives and FAO interventions, many developing countries thinking and implementing the practices that reduce GHGs emissions in order to mitigate climate change. Higher subsidy to input, very often lead to environmental degradation; so must be discouraged in developed as well as developing countries in perspective to climate protection. The actual conflict arises because the effect of the subsidy on crop production and economic benefit are short term in nature and easily identifiable and quantified by the people; whereas the effect of CSA practices on the environment and sustainable crop productivity is long-term and difficult to quantify in monetary term and not easily understood by common people. So, naturally, common people/farmers prefer "subsidy" to inputs than that of "incentive to CSA practices". This is the trade-off.

There are some good examples of "incentives to CSA practices" which were successful as CSA policy. Those are mentioned below:

(i) The incentive for using sugarcane trashes for bio-ethanol production in Brazil, a tropical humid country is a climate-smart policy.

(ii) Surely, the use of bio-energy reduces GHGs emissions and therefore technology for sustainable production of bio-energy is an efficient CSA practice. Therefore, support to grow crops for bio-energy either in the form of "subsidy" or "incentive to CSA practice" is an opportunity to mitigate climate change by reducing CO_2 emissions. Incentives for growing bio-energy crop in tropical humid region was found effective than the temperate region as any practice/technology become climate-smart depends on the ecosystem, enabling environment and existence of alternative land-use options.

(iii) There is huge potential to use of crop residue (rice and wheat straw) and woody biomass for bio-ethanol/bio-energy production in south-east Asia. Putting incentives for residue management could be a part of the CSA policy. Few countries in those regions already start thinking on that line.

(iv) The United States of America had subsidized production of maize for bio-energy production; however, the support measures were recently phased out because of competition of land for food crops.

(v) The countries of northern-Europe subsidized rapeseed for bio-energy and make it a part of their broader energy policies. However, there is debate about whether these measures are climate-smart or not in temperate ecosystems.

(vi) An effective CSA policy introduced by Vietnam government was "five reductions, one must". The policy spoke about five reductions, viz., reduction in seed to be planted, amount of water to be used, the amount of fertilizers to be applied, amount of pesticides to be used, and loss of produce after harvest. It was emphasized "one must" thing that is good quality seeds should be used. This policy has achieved remarkable results in terms of reduction in the use of fertilizers, water, and pesticides (around 50%) and increased the yield of about 15–20%.

Subsidizing inputs (fertilizers, water, and pesticide) and energy in agriculture may lead to an unsustainable system which is not "climate-smart". However, in resource-poor countries, subsidies to input up to a certain level improve crop productivity; maintain soil health and carbon sequestration. But, it is short or middle-term affairs; if it continues for a long time it would harm the environment and natural resources. Subsidies to agricultural inputs in "resources-rich" countries often lead to overuse or misuse of resources leading to unsustainable practices and cause land and environmental degradation which are not at all "climate-smart" (Fig. 11.1). Subsidies in terms of "support-price" to marketable products in developing countries ensure social and food security and could be considered as the "climate-smart" approach. Overall, subsidies on principle do not support the essence of climate-smart agriculture. However, reforming the pattern of subsidies is difficult as it is based on various socio-economic and political issues that are highly variable among the countries [3, 15].

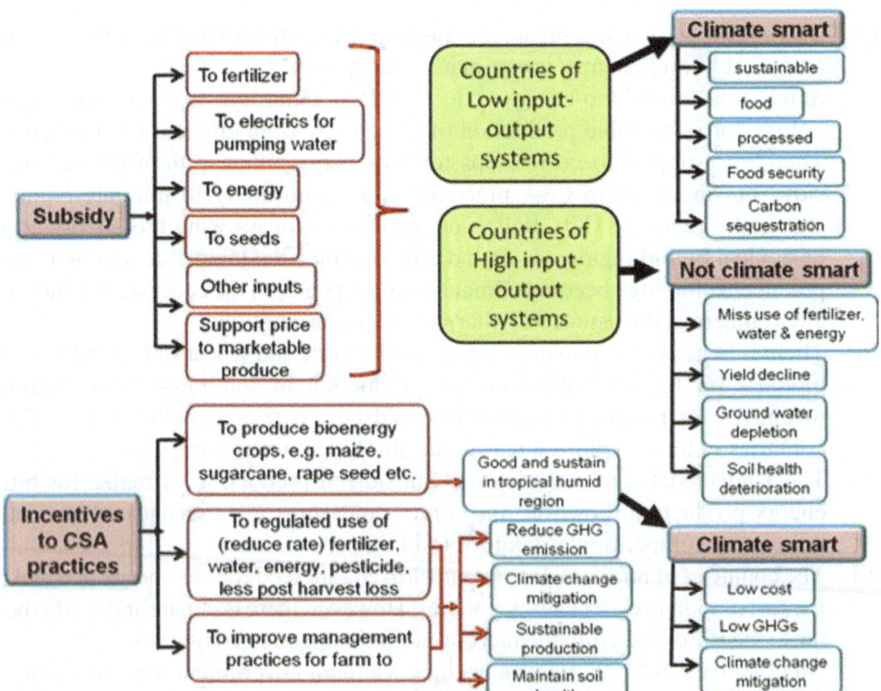

Fig. 11.1 Trade-Off between "subsidy" and "incentives to CSA practices"-conceptual model

"Incentives to CSA practices" must be encouraged in the CSA policy to promote environmental security for the long term. Incentives should be given to regulated use of fertilizers and water rather than the amount of that input used. It should be given to cultivating bio-energy crops in tropical humid regions keeping in mind that there should not be any competition for land for food and bio-energy crops. Incentives must be given for constructing and maintaining of rainwater harvesting structure. The water distribution system may be put in people's participation mode and pricing may be introduced for water use or else incentives should be given to those who increase their water and fertilizer use efficiency. Farmers should be paid for providing ecosystem services as CSA incentives and that should be integrated into CSA policy itself. Incentives also could be given to those who minimize the post-harvest losses.

The quantification and proper pricing of ecosystem services, carbon sequestration, water/energy saving, conservation of soil, and biodiversity which are long-term phenomena is the challenge. However, those services and intangible benefits have to be quantified and priced, and then only the CSA policies would be attractive and easily acceptable. Therefore, long-term finance is essential for effective CSA policies, and for that only mainstreaming the CSA practices into core national policies of the country is so important.

11.1.5 How to Make CSA Policies Effective and Acceptable

Three features/strategies can make the CSA effective and more acceptable at the national level irrespective of "resource-rich" or "resource-poor" countries. Those are (i) interlinked productivity-adaptation-mitigation to climate change through incentives; (ii) priced the ecosystem services derived from CSA practices; and (iii) regulation for CSA-adoption (Fig. 11.2).

The CSA policies which linked productivity with adaptation and mitigation to climate change and make provision to incentives for holistic approach have

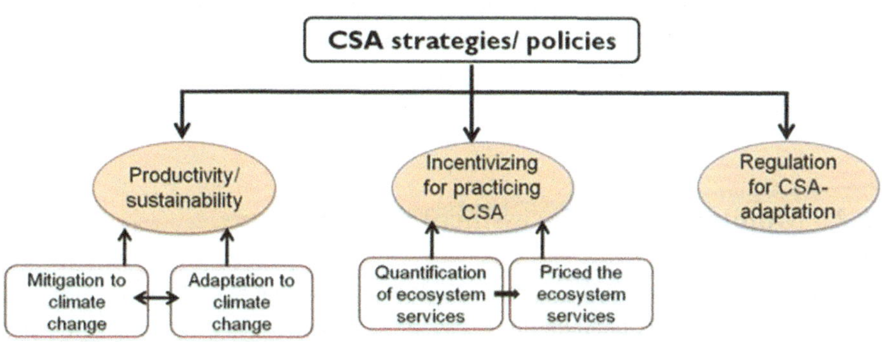

Fig. 11.2 Strategizing the CSA policies

got success. We can quote here, the policy or program of "Watershed Management through People Participation." In those program, higher crop production was achieved through assured water supply from rainwater harvesting (RWH) and supplemental irrigation. At the same time, it checks soil carbon erosion and energy use for irrigation which helps to mitigate GHGs emissions. So, the incentives and support should be given for making RWH structure and irrigation/drainage channels on a watershed basis through farmers' participation that makes the practices people friendly and enhance adaptation. For example, Karnataka Watershed Project in India (covering 0.5 m ha watershed area; 1.8 million people) and the Mid-Himalayan watershed project which ensured people participation; implementation and maintenance of soil and water conservation measures; increased system productivity at farm level; income generation and forest protection. Pricing of ecosystem services (both tangible and intangible) generated due to practicing of CSA is an important pillar of CSA policies. Benefits gain in short-, middle- as well as long term must be quantified, valued, and priced. And the reward or incentives for those should be given either through cash or kind via reliable transparent systems. It would motivate the farmers or stakeholders to follow CSA practices at the same time compensate for the apparent losses that may cause due to loss of subsidies. Financing must be encouraged to provide incentives to those who are practicing CSA practices rather than blanket subsidies to input-use to agriculture. In this connection, farmers, local agencies, NGOs, and the government should work together patiently as most of CSA approaches are long term in nature. Carbon finance that aims at climate change mitigation is a good example of incentives. So, we see that incentives could be in an indirect way, like through laws and regulations, or direct way, like cash subsidies, non-cash subsidies, including technical assistance, etc. However, the majority of successful CSA program often followed a mix of policy support with various sources of national and international funding. The third and important strategy is the regulations for the adoption of CSA. National regulation must be there even if the CSA practices are widely adopted to safeguard food security and public interest at the local level. For example, carbon-credit, production of bio-energy crops, the export of high-valued crops from higher production zone to distress-zone may be encouraged without compromising the national food security and public demand for food and livelihood. An excellent example of that, Ukraine banned (for short term) its wheat export during 2010-draught in order to protect their people's demand and maintaining food security of the country; even, it compromised its reputation in the world market on wheat export.

11.2 Monitoring and Assessment of CSA

Assessment and monitoring of CSA practices as such and or CSA policies as a whole is not easy. It is challenging because the impacts of CSA are mostly long term and related to environmental issues. Moreover, the long-term beneficial effects of CSA, many times cross the local and national boundaries as those are global

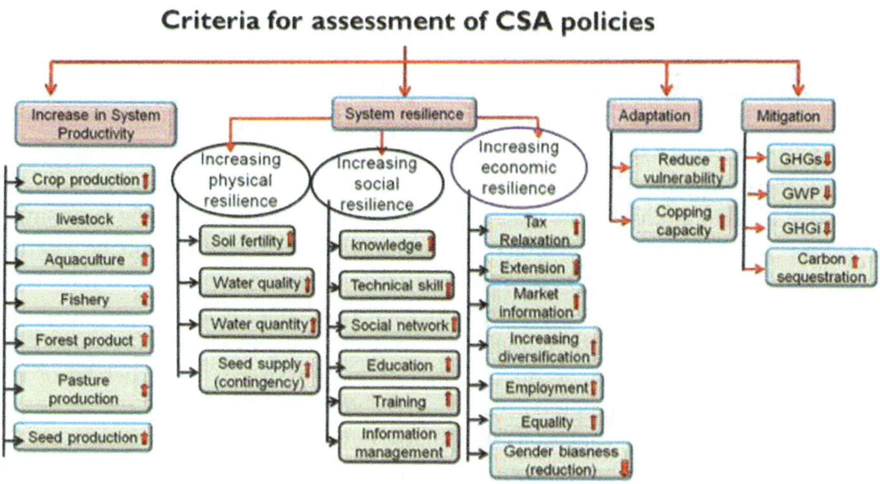

Fig. 11.3 Multi-sector-assessment or monitoring criteria of CSA policies

in nature (climate/environment is global). There are several facets of beneficial effects of CSA in terms of increasing resilience, enhancing the productivity of the system, and mitigating GHGs emissions and climate change. Therefore, evaluation/assessment/monitoring criteria of CSA/CSA policies must be multidimensional (on time frame) covering multi-sector and multi-locations. The assessment of CSA practices must be done in a short-, medium- and long-term time frame. The monitoring of CSA should also be linked with existing assessment systems for climate change development in agriculture. The multi-criteria evaluation has to be performed on the basis of four major aspects. Those are, sustainable productivity, increasing resilience, adaptation, and reduction of vulnerability and mitigation of GHGs emissions and or climate change (Fig. 11.3).

The sustainable productivity of the systems includes crop, livestock, pastures, rangelands, aquaculture, fisheries, forestry. Quantitative as well as qualitative measurements are required in those aspects. The assessment of productivity may be in short- as well as a medium-term basis. Productivity assessment should follow the system approach rather than an individual/single crop approach. Some indexes, like sustainable yield index, partial productivity index, etc., could be used for evaluation instead of a simple yield of crops. The second broad criteria of assessment of CSA are resilience. Resilience can be physical, social, and economical. Generally, the resilience of the system is measured in mid- and long-term time frames. Enhancement of physical resilience of the system is measured through water availability (higher quality and good quantity), improvement in soil fertility/health, consistent supply of good quality of seed in need/stress situations. System resilience refers to the rapid come back of the system in its original state after it was subjected to stress. Climate resilience of a system is specifically referred to bounce back to its original states which had distorted due to climatic stresses/vagaries. Economic resilience of the systems could be judged by the indicators, like, the betterment of public health, education, livelihood, income, and employment. It is also assessed through the increment of income-diversification, gender-equity, healthy market-information, etc. The increase of social-resilience of the system is depicted through the enhancement of knowledge, technical skill, social-networking, and the way general people/farmers manage the information (particularly, weather and market). Reduction of vulnerability due to adaptation of CSA practices is an important indicator of the assessment of CSA. Adaptation enhancement or reduction of vulnerability could be alternatively used for judging the performance of CSA. However, it is also a long-term phenomenon. Clubbing the mitigation strategy with adaptation and resilience makes the CSA approach novel and unique. Therefore, the mitigation has to be quantified in order to assess the CSA. Reducing the GHGs emissions (t CO_2 per ha or per year) and or mitigating the global warming potential (GWP) (t CO_2 per ha or per year) are key criteria for evaluation. Sustainable development in terms of agricultural production and agricultural economics is also an essential component of CSA, so the mitigation can also be judged in terms of GHGi (Greenhouse gases intensity, i.e., GHGs emissions per unit production) (Fig. 11.3).

We know that assessing climate change adaptation and resilience is a debatable issue. Therefore, a single approach/criteria are not enough and also not scientific to monitor CSA. As, for example, adaptation increases the coping capacity to climate change vagaries, and at the same time, vulnerability hinders adaptation. The adaptation and vulnerability both are driven by several physical, social, and economic factors that are interlinked and sometimes complementary in nature. Those factors are location-specific and time-dependent also. So, different levels of monitoring systems are required with proper synergy and networking. Decentralization of assessment of physical criteria and centralization on incentives for rewarding the stakeholders who are following the CSA practices is the essence of CSA policy. Mid-term evaluation of the impacts of CSA also should be considered as most of the CSA practices results in environmental and social benefit in long-run. Criteria for accounting the saving of natural resources (soil, water, air, biodiversity, ecological-balance, etc.) due to the adoption of CSA must be considered for assessment and future funding [19].

11.3 Financing in CSA

The success of the CSA policy depends on adequate and logical funding. Financing the CSA is the key to mainstreaming it into core national policy. We understood from the previous chapters of this book and the previous sections of this chapter, CSA is an integrated approach and the benefits of this would generally of long term that deals with environmental, ecological, and social issues. Therefore, investment in CSA also should be integrated and long term considering the future negative effect of climate change. In easy terms, the adaptation of CSA may require more investment in future adverse climate change consequences. At the same time, mitigation may reduce the negative impact of climate change in the near future; in that case, investment requirements may be less. So, financing to CSA should be based on future climate change predictions by keeping the provisions of mid-term correction. In agriculture, the majority of investment has been done internationally as well as nationally on the basis of productivity increase and food security goal. Three aspects are generally considered, viz., (i) poverty reduction, (ii) food security, and (iii) economic development. Financing on those three aspects was more prominent in "resource-poor" developing countries. Investment in climate and environmental safety was thought to be luxury in those countries even 4–5 decades ago. There were also existed conflicts and lack of consensus on investment to climate-safety or development/or both among the world leaders and countries having different economic-set up.

However, climate change is a fact. It has been evident through the occurrence of enhanced frequency of drought, floods, tropical cyclones, melting of glaciers, and consistent sea-level rise in the last century. More so, climate change is projected to have severe negative effects on agriculture in the near future. Agriculture and farming communities of resource-poor countries are more vulnerable to climate change. Therefore, today's investment plan must incorporate projected future impacts of climate change on agriculture. So there are seven (7) things which must be considered for CSA financing and those are listed below:

(i) Integration of CSA finances/investment into core National Planning.
(ii) Direct or indirect "incentives for practicing CSA practices" (after proper multi-criteria assessment) and provision of money for that should be the part of financial-planning.
(iii) Current investment has to consider future climate threats (i.e., cost of dealing with future threats projected due to climate change) for adapting CSA practices.
(iv) Financing also should consider mitigation potential which may reduce the future climate change threats. So, mid-term correction of financial allocation should be in the card for the proper utilization of money and resources.
(v) Unlike the previous climate-financing which are mostly on mitigation aspect in energy and industrial sectors; current CSA financing should consider mitigation and adaptation cost together in energy, industry, and agricultural sector, simultaneously.
(vi) National funding must be boosted up by international funding for higher CSA adaptation as climate change and environmental safety are global phenomena although that has to be tackled locally.
(vii) And the "seventh" point is that the private, as well as public sector investment, should go hand by hand to reduce the vulnerability of farmers and protection of our environment.

11.3.1 Challenges and Opportunities for CSA Financing

There are a lot of opportunities which emerge currently on CSA financing, although some challenges still exist. The challenges and opportunities of CSA financing are presented below in tabular form for easy understanding (Table 11.1).

Table 11.1 Challenges and opportunities for CSA funding/finances

Challenges	Opportunities
i. Basically low investments in agriculture; more so, on remunerative activities	i. A lot of international climate finances are available like UNFCC, UN, GFC GEP
ii. Mass investors are farmers, foresters, hoarders, fishers themselves (Private investors and many a time small investor)	ii. Current emphasis is given on public investment on CSA
iii. Mostly concern about sustainable agricultural productivity and food security and not about profit. So, private investors generally do not found it lucrative	iii. After "Rio Summit", "Paris-Agreement", and "IPCC-AR-6" documentation; the emphasis has been given to financing issue of CSA in the interim report of IPCC in 2018, 2019,
iv. All-climate sector financing is on mitigation and in energy and industrial sectors only	iv. The new form of GFC (Green Climate Fund) has provided the opportunity to fund on mid-term CSA adaptation strategies
v. Traditionally, the majority of funding on either mitigation or adaptation	v. Global environmental facilities move toward synergies of adaptation-mitigation; could directly financing the CSA activities
vi. The voluntary carbon market is small; because of high transaction costs and lack of methodologies for evidence and quantification of carbon save or gain	vi. Agricultural carbon-credit; methane avoidance from manure management; fertilizer use efficiency based financing; forest certification is coming up
vii. Lack of "CSA-evidence" based institutional facilities for certification of C-credit and energy saving in developing countries	vii. Carbon marketing could be a good source of finance for CSA

The adaptation of CSA is a challenge in the world where 65–70% of rural people are food insecure and indirectly or directly dependent on agriculture as a sole source of income [1, 8]. Further, those "resource-poor" sections of global-society resided primarily on South Asia and sub-Saharan Africa are more vulnerable to climate change [11] and going to be affected worse in coming 20–30 years. So, support for sustainable agricultural growth is the primary concern of those regions. And the adaptation of CSA practices in general and long-term strategies, in particular, is very difficult in such regions. Investment by international and national agencies that directly target poverty alleviation and with in-built compatible CSA practices could be useful [1, 2, 4, 9]. Poverty alleviation of rural vulnerable people with agricultural growth for food security itself is a key adaptation strategy of CSA which increases the resilience of the system as a whole. At the same time, we should also address the adaptation deficit, incremental change of both quantity (physical aspects), and quality of life and livelihood. In this goal investment is the engine and it must be adequate to start and maintain growth.

The action for adaptation of CSA can be looked into three perspectives. The first one is to remove the adaptation deficit. The second one could be the incremental changes in the physical aspects of adaptation for climate-proofing. The third one is to bring the transformational changes by qualitative modification of the human-behavior

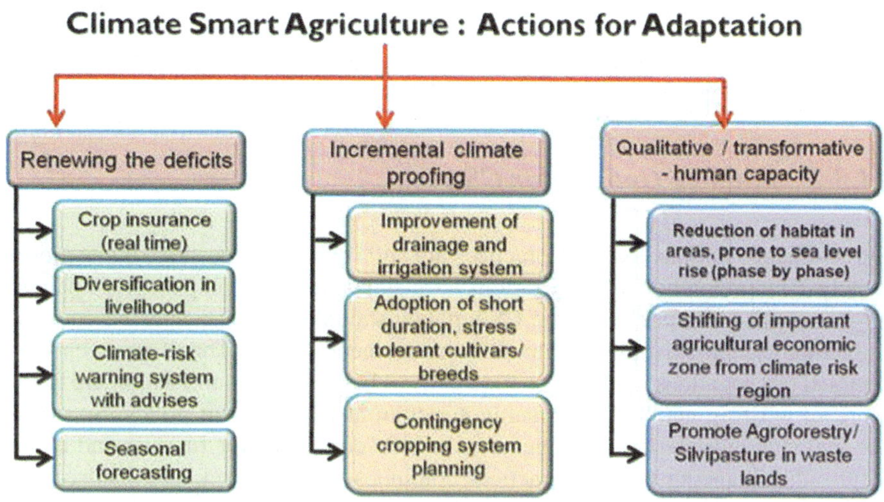

Fig. 11.4 Avenues in which financer look-into for higher adaptation of CSA

through capacity building (Fig. 11.4). Real-time crop-insurance, diversification of livelihood options, precise weather forecasting on the seasonal or intra-seasonal basis in agricultural crops, an effective warning system for disaster management with efficient agro-advisory are the important activities which could reduce the adaptation deficit of CSA. The incremental physical change of landscape, management practices for climate-proofing facilitates adaptation. Improvement of drainage, as well as irrigation network on watershed basis in hilly and undulating areas and increasing water use efficiency in the plain by reducing conveyance losses can promote adaptation. Introduction of abiotic-stress (drought, floods, salinity, submergence, heat stress, etc.) tolerant cultivars in climate-stress zones and also promotion of short duration varieties for avoiding the water and temperature stress helps in higher adaptation. In this context, contingent crop planning is useful which helps in building the confidence of farmers toward the government-CSA policies and on implementing agencies. Transformational changes are important for better adaptation. For example, intensive agricultural activities must be relocated to new/safer places from the region where sea-level rise is predicted and salinity enhancement is on cards. Similarly, an economically important zone of agriculture has to be shifted to relatively safe places based on future climate-risk scenarios. Capacity building of human skill should emphasize simultaneously for better adaptation. A low-input agricultural systems, like silvipastoral, pastoral, agroforestry, etc., should be promoted in predicted climate-risk regions for easy adaptation and reducing the vulnerability (Fig. 11.4).

In nutshell, CSA financing should be matched with national sustainable agricultural development goals. Investment priorities must consider the present economic condition of the country; future climate change risk; vulnerability of the people engaged in agriculture; mitigation potential of GHGs emissions/climate change in mid-, long-term scenarios; the pace of technological development; and the challenges to sustaining food and nutritional security in burgeoning global population.

11.3.2 Finance Options for CSA

Direct finance options for CSA practices are still in the budding stage but exploring slowly. Few international organizations have shown interest and paths for financing CSA activities. On principle, the CSA policy should consider a specific country's vulnerability and existing climate-risk management system. Therefore, even international organizations are ready to fund CSA; it has to be executed within the framework of existing national sustainable agricultural development policy. So, unless the national and international agencies develop consensus for funding and expenditure, the financing to CSA practices continues to be limited. The landscape of CSA financing has complex features having various objectives and difficult eligibility criteria. The international organization like, UNFCC (United Nations Framework Convention on Climate Change) has some direct financing options under GEF (Green energy fund) which has two components, viz., SCCF (Special climate change fund), and LDCF (Least developed countries fund). The CBD and UNCCD mostly invest in biodiversity restoration and restoration of land degradation; these funds could be indirectly merged with CSA finance with a long-term goal. The GEF trust fund is primarily for climate change mitigation, although recently it is promoting CSA practices worldwide in the land-use change and agriculture sector. There are possibilities of getting indirect finance from CCA financing to the developing countries who agreed with the Kyoto Protocol. National level food security program includes CSA practice that can obtain funds from the UN organization and or MDBs (Multilateral Development Banks). The country level public financing in collaboration with private funding agencies would be the future of CSA financing that needs to be strengthened. In those investments "incentives for providing ecosystem services" is the key. The voluntary carbon market is the potential finance agency for CSA but methodologies for carbon-credit assessment have to be standardized with a reduction of transportation cost. Private sector financing has been starting but a long way to go; however, we must consider agriculture is by and large a private activity with individual ownership, so unless CSA practices would be remunerative those would not be adapted in large scale [16].

11.3.3 Carbon-Credit

"Carbon-credit" in agriculture and allied sectors has the potential to generate sufficient finance for CSA. It might be the future biggest financing opportunity to CSA practices. However, till now it is in the baby stage and far behind from the industrial and energy sectors. In the energy and industrial sector, there have been well-defined methodologies for quantifying "reduction-emission" and "carbon-saving." They also have transparent well-evidenced certified agencies, who provided the "Emission-Reduction-Certificates" that are accepted globally. But in agriculture in particular and "land-use change, forestry, and agriculture" in general, there are lot of uncertainties. Global debate still persists from the last 3 decades, whether emission from biogenic sources (agriculture, plants, forestry, etc.) should be considered for "carbon-credit" compliance. Sources and sink capacity of carbon in agriculture and cropland are not restricted to the field alone, it reaches far behind from farm-gate to end-consumer via many short or long-chain processes (threshing, processing, packaging, storage, transportation, export-import, consumption, bio-availability, recycling of bio-waste, etc.). Therefore, methodologies for accounting carbon-saving a certain agricultural product is difficult because there are many leakages. Certification on "emission reduction" is also difficult because putting boundaries to a certain point of food-chain is not easy and often debatable. Further, the majority of "resource-poor" developing countries are primarily dependent on agriculture. Putting pricing on "carbon-emission-reduction" without proper legislation (which is country-dependent and also some time international in nature) in agriculture may exert higher pressure on the developing country both on the economy and food security. There would be also a higher opportunity costs for making agriculture to "carbon market-ready" from "conventional climate change efforts" in most of the developing countries. Moreover, the majority of farmers in "resource-poor" countries are small and marginal having small landholding. Compliance for "carbon-credit" by combining a large number of fragmented land with different ownership is a difficult task, and so often failed to get "emission-reduction-certificate" for compliance and trading. And in those cases, the transaction cost is usually high and hardly gets any profit from "carbon-trading." Apart from that, the lucrative incentives for carbon-trading by growing bio-energy crops (higher carbon sink potential) have already shown a threat to food production and food security in developed as well as developing countries. So, the pricing of GHGs reduction and "carbon-saving" in agriculture has been challenging and also a debatable issue. Moreover, due to above-mentioned controversies, doubts still exist in the mind of world leaders, researchers, policymakers, and farming communities as a whole, the "Certified Emission Reductions" and "Carbon-Trading" from cropland, soil management in agricultural land, and pastures/grassland was not considered "Clean Development Mechanism."

Despite all these challenges, we believe that "carbon-credit" in specific and "carbon markets" in general could be a huge source of funding for CSA both nationally and internationally in the future. We can make the challenges into opportunity by introducing innovative methodologies for estimating carbon-saving and or GHGs emission reduction quickly and precisely in agriculture. Reliable tools for measuring carbon-saving could eliminate the un-stability of the global carbon market. Conflict resolution of developing and developed countries over energy use and carbon-saving (unsustainable development vs. environment safety) looking into the seriousness of global warming and climate change consequences (negative impact) can sow a new path for CSA financing. Inclusion of agricultural carbon-credit in "Clean Development Mechanism" and "European Union's Emissions Trading System" protecting the interest of "resource-poor" countries with proper legislation and regulation may help in CSA financing. The incentive to methane avoidance in manure management is a good example which is being successful. Overall, until now the agriculture sector is lagging in terms of carbon-credit and carbon-financing than the industry and energy sector. Working knowledge, experiences, and precise methodologies are necessary to overcome the barriers [12]. However, voluntary carbon markets and some agencies like "Verified Carbon Standard," "American Carbon Registry' that consistently promote carbon credits from land-use change (LUC), land use (LU), and forestry activities, throw some hope in this regard [18]. They allow for carbon credits from CSA activities also. But even financing from those including REDD to CSA is still very meager. The newly emerging carbon market (voluntary) in China, using their indigenous "carbon standard" (Panda Standard) for climate change and GHGs mitigation from agriculture in encouraging and can be promoted in other parts of the world with proper validation, regulation and legislation [5, 17].

11.4 Summary and Messages

- The basic principle of "sustainable agricultural development" and "climate-smart agriculture" although the same, the CSA requires little more specific policies on mitigation, resilience development, and adaptation.
- The CSA strategies must be a part of core national agricultural policies which should consider the natural-resource potential and existing financial situation of the respective countries.
- Mainstreaming of CSA policy should be based on the reliable national database, the vulnerability of people, evidence and prediction of climate change, current social and political situation of the country, barriers to adaptation, and potential of the country to cope up future environmental and economic challenges.
- The CSA policies need to be multi-sectoral, integrated that includes food security, poverty alleviation, economic growth, mitigation of GHGs emissions, and system-resilience together. Policy-framework may be centrally developed but executed locally (decentralized manner) keeping all the sister government and private agencies taken into confidence. The policies also should have short-, mid-, and long-term objectives and milestones and deliverables.
- As the CSA deals with both economic and environmental issues, so transparent legal, regulation and evaluation framework is essential. The monitoring, regulation, and support mechanism must be institutionalized and responsibilities are fixed to keep the proper coordination among the cross-sectors functionaries.
- A support system comprising of both "subsidy" and "incentives to CSA practices" should be balanced based on the country's socio-political perspectives. Regulations framework should promote incentives for following CSA practices than a subsidy to input only keeping in mind the long-term sustainable growth of the economy as well and environment-protection.

- Behavioral changes of the public are necessary by advocating the thought that investment in CSA would reduce food insecurity, enhance productivity, develop system-resilience, mitigate climate change, and GHG emissions in long-run.
- Financing to ecosystem services and carbon-saving through incentives to all stakeholders practicing CSA is the key to CSA adaptation. It needs transparent methodologies for pricing both tangible and intangible benefits generated from CSA.
- Present-day, in the transitional period of CSA, adaptation needs international, national, public as well as private financing not only to agricultural input (for agriculture) but also price-support to technology innovation, increasing nutrient and water use efficiency, regulations, infrastructure development, social protection, networking, and transportation.
- The international and national financing organizations including banks, NGOs, private organizations, social-welfare groups, etc. should join together for long-term investment for promoting CSA.

References

1. Bhattacharyya P, Munda S, Dash PK (2019) Climate change and greenhouse gas emission. New India Publishing Agency, New Delhi, 110088; ISBN: 978-93-895-7175-2. pp 170+25
2. Bhattacharyya P, Roy KS, Nayak AK (2016) Greenhouse gas emission from agriculture monitoring, quantification and mitigation. Narendra Publishing House, Delhi India. ISBN: 978-93-84337-00-0, pp 193
3. Chandra A, McNamara KE, Dargusch P (2018) Climate-smart agriculture: perspectives and framings. Clim Policy 18(4):526–541
4. De Janvry A, Sadoulet E (2010) Agricultural growth and poverty reduction: additional evidence. World Bank Res Obs 25(1):1–20
5. Dhar S, Pathak M, Shukla PR (2017) Low carbon city: a guidebook for city planners and practitioners. Low Carbon Transport in India Project.
6. FAO (2010) "Climate-smart" agriculture: policies, practices and financing for food security. Adaptation and Mitigation, Food and Agricultural Organisation, Rome.
7. FAO (2012) Agriculture Organization of the United Nations. 2012. FAO statistical yearbook.
8. FAO, IFAD, UNCTAD, The World Bank (2017) Principles for responsible agricultural investment that respects rights, livelihoods and resources. In: Food and agriculture organization of the united nations, the international fund for agricultural development, the united nations conference on trade and development and the world bank, Rome and Washington, DC
9. FAO (2008) The state of food insecurity in the world: high food prices and food security-threats and opportunities. pp 1–2 ftp://ftp.fao.org/docrep/fao/011/i0291e/i0291e00a.pdf
10. Food and Agriculture Organization of the United Nations (FAO) (2012) "Climate-smart" agriculture: policies, practices and financing for food security, adaptation and mitigation; food and agriculture organization of the united nations (FAO), Rome, Italy
11. HLPE (2012) Social protection for food security. A report by the high-level panel of experts on food security and nutrition of the committee on world food security
12. IMF O (2016), UN and World Bank (2011) Supporting the development of more effective tax systems. Report to the G-20 Development Working Group
13. International Institute for Sustainable Development, Mann HL (2005) IISD model international agreement on investment for sustainable development

14. IPCC (2018) Summary for policymakers. In: Global warming of 1.5°C. An IPCC Special Report on the impacts of global warming of 1.5°C above preindustrial levels and related global greenhouse gas emission pathways, in the context of strengthening the global response to the threat of climate change, sustainable development, and efforts to eradicate poverty [Masson-Delmotte V, Zhai P, Pörtner H-O, Roberts D, Skea J, Shukla PR, Pirani A, Moufouma-Okia W, Péan C, Pidcock R, Connors S, Matthews JBR, Chen Y, Zhou X, Gomis MI, Lonnoy E, Maycock T, Tignor M, Waterfield T (eds)]. World Meteorological Organization, Geneva, Switzerland, 32 pp
15. Khatri-Chhetri A, Aggarwal PK, Joshi PK, Vyas S (2017) Farmers' prioritization of climate-smart agriculture (CSA) technologies. Agric Syst 151:184–191
16. Lipper L, Zilberman D (2018) A short history of the evolution of the climate smart agriculture approach and its links to climate change and sustainable agriculture debates. In: Climate smart agriculture. Springer, Cham, pp 13–30
17. Pipeline, UNEP Risø. "UNEP Risø CDM/JI Pipeline." UNEP Risø Centre, Roskilde, Denmark (www.cdmpipeline.org) (2011)
18. PWC and the Rockefeller Foundation, Gledhill R, Herweijer C, Hamza-Goodacre D, Grant J, Webb C, Steege J (2011) Agricultural carbon markets: opportunities and challenges for Sub-Saharan Africa
19. Thornton PK, Lipper L (2014) How does climate change alter agricultural strategies to support food security?, vol 1340. International Food Policy Research Institute
20. United Nations Environment Programme (2017) The Emissions Gap Report 2017: A UN Environment Synthesis Report. The Emissions Gap Report 2017, United Nations Environment Programme (UNEP), Nairobi, Kenya, 1–86 pp

Chapter 12
Messages and Way Forward

12.1 Introduction

The challenges to feed around 9 million world population in 2050 by producing 60% more food is a daunting task. The challenges are more difficult by increasing constraints of freshwater availability for crop and livestock and predicted climate change risk. The concern about climate change and its impact on agriculture have been recognized since the first earth summit in 1992 and again reiterated in "COP 23" in 2017. It was emphasized that both opportunities and constraints on food security, system-resilience, adaptation, and mitigation should be addressed by SBI and SBSTA with synergy [1]. The encouragement must be given on "technological innovation," "investment," and "site-specific action" to "climate-smart agriculture" (CSA) to address the issues of climate change and food security jointly.

12.2 Eight-Point Agenda to Look Forward

12.2.1 Needs Large Shifts in Land-Use Patterns and Crop Choice

Land acquisitions on a proper scale based on their nature and potential use are important in the context of tenure security and climate change. Effective land-tenure systems have a significant impact on climate change mitigation and adaptation. However, land acquisition and further putting of those lands into an alternate resilient cropping system should be done carefully with due consideration to socio-economic perspective and legal contexts. For example, communal tenure for forest and range-lands would be beneficial for profitability, carbon sequestration, and climate change mitigation, but strong public support and coordination of public administration are

© The Editor(s) (if applicable) and The Author(s), under exclusive license to Springer Nature Singapore Pte Ltd. 2020

P. Bhattacharyya et al., *Climate Smart Agriculture*, Green Energy and Technology,
https://doi.org/10.1007/978-981-15-9132-7_12

necessary for that. Further, the land currently not used for crops, but have the potential to fetch higher yield in predicted climate change scenarios needs to be targeted for shifting/changing of land-use pattern and crop choice [2]. Actually, a coherent land and climate policy would be required which not only reduces the risks, minimize losses but also give logical returns to the investment. Therefore, suitable and coordinated "land-climate" policy-frame work at different levels, scales, and sectors would address the equity and fetches the benefits with respect to "climate risks-reduction" and "improvement in food security."

12.2.2 Mapping of Vulnerable Zones and Climate-Hazard-Risk

The "flood-zone mapping"; "drought-zone mapping"; "salinity-zone mapping"; "zones of water scarcity" (predicted); "zone of weather-extreme events" (predicted); "multi-hazard zone mapping" (predicted) are the advanced products in remote sensing and GIS platforms that need to be utilized properly for CSA policy-making for future. The mapping and risk assessment should go hand by hand considering the future climate change scenarios. The map-based site-specific CSA planning needs to be merged with a people-participatory action plan through government investments and international aid.

12.2.3 Change in CSA Research Aptitude

Breeding for abiotic multiple stress tolerance cultivars for crop/and breed for livestock will remain important in the CSA research agenda. However, the research aptitude toward CSA should be delivery oriented rather than discovery or testing based. The CSA research must be action-research. Future, research of CSA should focus on adaptive practices and adaptive governance at the local scale in the social-ecological system having "long-term human-provisioning." The adaptive research must be encouraged on local, national, and international scales with the systematic integration of information that logically frame the post-disciplinary science agendas. The shifting of "agricultural-research" toward development trajectories having higher certainty is actually "climate-smart" research. Therefore, a transformation of global agricultural systems with low emissions portfolios would be the future research goal of CSA.

12.2.4 Safe Operating Spaces in the Context of Climate Change

The "climate-smart agriculture" is a sustainable way to address food security, adaptation, and mitigation to climate change together. But researchers from one school of thought stated that the "three dimensions of CSA are poorly understood". They thought any "improved agricultural practice" could be called "climate-smart" and there is a lack of clarity between CSA's "concept" and "practice." However, the CSA conceptualizes in a holistic way that could bring all the stakeholders closer to "safe operating spaces" in agricultural across the temporal and spatial scales. Therefore, efficient agricultural management practices that brought us nearer to "safe operating spaces" would definitely require transformations in governance rather than small incremental changes including effective utilization of natural resources, congenial social, political, and economic conditions. Further, quantitative indicators to assess the long-term "safe operating spaces" in relation to climate change are important for achieving the "transformational changes." So, the future efforts should be given on the prioritization of key areas, like, effective governance of social-ecological systems; integration of social-ecological and socio-economic indicators; integration of knowledge, data, tools, and models for real-time decision-support systems; and transparent policy dialogues on local to international scales to create the "safe operating spaces."

12.2.5 Needs Transformational Changes

Effective CSA practices should be promoted both in developing and developed countries as "transformational changes" not as incremental adaptation measures for meeting the challenges of "food security and climate change." In this context, as discussed in previous chapters of this book the ecosystem approach should be adopted, with inter-sectoral integration and cooperation, and could be executed at the landscape level to address the climate change threats. In general, CSA needs investment for innovative technology development and adaptation, development of climate change-stress-tolerant varieties/breeds, strengthening the institutional capacity and skills, and to finance the small farmers to make the transition to CSA. The CSA policies should be inclusive and integrated with national level core agricultural policies with the assured funding from international, national, and regional level organizations. However, the future of CSA depends on how much synergies among public and private financing sources could be achieved at fast-track mode on a sector-specific basis and socio-political support availed for agriculture.

The synergies between improved food security, adaptation, and mitigation in the CSA platform for marginal and small farmers should be critically identified, developed, validated, and disseminated. As the majority of "resource-poor" countries preferred adaptation to climate change as the priority for availing future food security but has reservations toward mitigation-measures/commitments; so finding the synergies is essential. After establishing synergies among adaptation, food security, and mitigation approaches the scaling up is necessary which essentially requires transformative changes in current CSA policies.

12.2.6 Needs Integrated Private and Public Sector Financing

Like all other innovative approaches, the CSA is also highly dependent on financing for its successful promotion and adaptation. Fortunately, the current international climate finance is in favor of CSA. But the small and fragmentation of the investments is a barrier to get benefits from multi-sectors agencies basically involved in CSA. Therefore, future financing to CSA should try to combine the "thematically separated channels" of funding sensibly and effectively. As, for example, private sector carbon-finance may combine with public financing to "forest degradation/competing land use" (both are addressing the climate change issues) in order to put more explicit emphasis on CSA activities. The Green Climate Fund and GEF-6 replenishment fund of UNFCC could be utilized in this context. However, for getting benefits from GCF financing for CSA the "resource-poor" countries have to develop effective monitoring systems; make available the good quality data; robust institutional-framework; technical skill and transparent legal frameworks which are prerequisite for international investment. Financer also needs priority setting and needs assessments at the national level (e.g., through NAPAs, "Nationally Appropriate Mitigation Actions", etc.) that provide a strong basis for country-specific measures. So, to start-up, the developing countries must keep their "fundamentals in place" to grab the opportunity of CSA financing from international organizations.

12.2.7 Changing the Policy-Portfolios

Traditional "input-use"-based subsidy is no longer sustainable and not advisable in CSA. The policy-portfolios need a paradigm shift from "input-subsidy" to "incentives" for ecological restoration and "insurance" for health and climate-disaster risk. The investment must be inclusive having social safety, contingent-financing, and people-resilience in cards. The flexible carbon credits, climate-risk assessment, safety nets, access to weather forecasting/warning systems, and real-time agro-advisory should be in-built in CSA policies.

The future of CSA adaptation greatly depends on how people would participate in policy-decision-making starting from identifying the problems, collecting the weather data, responding in the early-warning system, facilitating and mediating soil/landscape conservation strategies, selecting the climate-resilient technologies to be adopted, choosing the indicators for evaluation or monitoring and more importantly in maintaining transparent financial flow from the investor to consumer.

Transparent methodologies to be developed for monitoring or assessing the CSA policies with multi-sector-criteria for interim corrections and modifications of strategies as most of the CSA practices are mid- or long-term in nature. In this context, adaptive governance for measuring progress toward goals is important. So the indicators with respect to people participation, land degradation neutrality, providing the ecosystem services, diversity restoration should be inclusive in evaluation framework. Further, an interactive, flexible, and adaptive mechanism for evaluating the risks, policy-instruments, and revising decision-support system are the portfolios that have to be integrated into future CSA policies.

12.2.8 Changes in Socio-Cultural and Political Outlooks

The positive political will and government support (both national and local level) toward "sustainable land-use change" keeping in view the climate-stabilization and triggering mitigation-efforts is the prerequisites of CSA adaptation. Polycentric hybrid governance of the private-public sector with equitable management of trade-offs, climate-risk, and opportunities would be required for CSA. To achieve that, empowering women and abolishing gender-bias among household asset sharing and food security would be ensured. The extensive presence of women in the majority of agricultural practices and land-based activities certainly open-up the opportunities to promote gender equality, reduce gender barriers, and mainstreaming the gender policies in CSA. Empowering women should be holistic by providing access to technology, information, training, extension, and as well as financing to them in a structured manner.

If we talk about the changes in socio-cultural outlook, then the "Indigenous-Local Knowledge (ILK)" plays an important role to combat climate change issues and food security at the local level. The ILK needs to be integrated with national CSA-framework to tackle the site-specific issues of climate stresses, technology adaptation, and ecosystem restoration. This approach is very effective as ILK is highly context-specific, multi-functional which take care of peoples' rights and social-behavior and also could be transmitted informally. Even, the ILK could be used for project-formulation; decision-making; and most importantly as an "entry-point" mechanism for "participatory-climate-action" (PCA).

References

1. Loboguerrero AM, Campbell BM, Cooper PJ, Hansen JW, Rosenstock T, Wollenberg E (2019) Food and earth systems: priorities for climate change adaptation and mitigation for agriculture and food systems. Sustainability 11(5):1372
2. Pugh TAM, Müller C, Elliott J, Deryng D, Folberth C, Olin S et al (2016) Climate analogues suggest limited potential for intensification of production on current croplands under climate change. Nat Commun 7(1):1–8